Soft Computing in Renewable Energy Technologies

This book addresses and disseminates state-of-the-art research and development in the applications of soft computing techniques for renewable energy systems. It covers topics such as solar energy, wind energy, and solar concentrator technologies, as well as building systems and power generation systems. In all these areas, applications of soft computing methods such as artificial neural networks, genetic algorithms, particle swarm optimization, cuckoo search, fuzzy logic, and a combination of these, called hybrid systems, are included. This book is a source for students interested in the fields of renewable energy and the application of soft computing. In addition, our book can be considered a reference for researchers and academics since it will include applications of soft computing in different renewable energy systems.

Soft Computing in Renewable Energy Technologies

Edited by Najib El Ouanjli, Mahmoud A. Mossa, Mariya Ouaissa, Sanjeevikumar Padmanaban, and Said Mahfoud

Department of Applied Physics, Faculty of Sciences and Technology, Hassan First University, Settat, Morocco

CRC Press is an imprint of the
Taylor & Francis Group, an **informa** business

First edition published 2025
by CRC Press
2385 NW Executive Center Drive, Suite 320, Boca Raton FL 33431

and by CRC Press
4 Park Square, Milton Park, Abingdon, Oxon, OX14 4RN

CRC Press is an imprint of Taylor & Francis Group, LLC

ISBN: 978-1-032-61191-4 (hbk)
ISBN: 978-1-032-61192-1 (pbk)
ISBN: 978-1-003-46246-0 (ebk)

DOI: 10.1201/9781003462460

Typeset in Sabon
by Apex CoVantage, LLC

Contents

Preface

Given the global energy demand, environmental impacts, economic requirements, and social concerns, electricity generation from renewable energy sources has become competitive with fossil fuels across many countries. Renewable energy sources such as wind, solar photovoltaic and thermal, hydroelectric, geothermal, and biomass are abundant and clean. Solar PV and wind energy are the most promising, with rapid growth, which are among the largest energy producers. The evolution of these sources and their integration in the network require an appropriate characterization of technologies and components of these systems for optimal performance under economic, technical, awareness and information, financial, and environmental barriers.

To ensure optimal performance of renewable energy systems, it is interesting to design control systems that emulate some functions performed by the human brain. Among these interesting functions are self-adaptation, learning, flexibility of operation, and planning in the presence of large uncertainties and with minimal information. Based on these aspects, soft computing techniques can be developed and applied to renewable energy technologies.

The book mainly deals with soft computing techniques for modeling, analysis, prediction of the performance, and control of renewable energy systems. This book will provide conceptual as well as practical knowledge about soft computing techniques used in renewable energy systems to the students, researchers, and academicians.

This book is organized into nine chapters in total.

Chapter 1 intends to study the brief introduction and past history relevant of renewable energy technologies.

Chapter 2 presents a study on issues and challenges of solar photovoltaic energy systems and soft computing applications to resolve these issues.

Chapter 3 introduces the tiki-taka algorithm for parameter identification of a three-diode photovoltaic solar module.

Chapter 4 presents a comparative analysis between the cuckoo search (CS) metaheuristic method and classic perturb and observe optimization method for PV System Under Uniform Condition.

Chapter 5 aims to present an overview of the solar concentrator technologies and their contribution in the development of CSP plant and analyze the possibility to combine two solar concentrators to optimize the energy production (electricity and heat).

Chapter 6 describes the various soft computing and artificial intelligence techniques and how these approaches have been applied to wind energy conversion systems.

Chapter 7 presents an innovative approach to optimize power extraction in wind energy conversion systems through an artificial neural network-based maximum power point tracking (MPPT) algorithm, whose objective is to obtain precise tracking of the maximum power peak despite varying wind speed.

Chapter 8 proposes an intelligent MPPT controller based on the African vulture optimization algorithm to improve the exploitation of wind energy.

Chapter 9 focuses mainly on controlling the speed of a switched reluctance motor for an electric ship propulsion system using fuzzy logic approach.

About the Editors

Najib El Ouanjli, PhD, is currently a professor in the Department of Applied Physics at the Faculty of Sciences and Techniques, Hassan First University, Settat, Morocco. He received his PhD degree in 2021 from Sidi Mohammed Ben Abdellah University, Higher School of Technology, Fez, Morocco and his master degree in 2015 from the Faculty of Sciences Dhar El Mahraz, Fez, Morocco. He was a professor of physics at the Science Ministry of Education from 2013 to 2021. His research interests are focused on renewable energy systems, electrical and electronics engineering, rotating electric machines, system modeling, control techniques, optimization techniques, fault diagnosis, power systems modelling, reliability engineering, grid integration, photovoltaics systems, smart grid, wind turbines, and solar energy. He has published several papers (international journals, book chapters, and conferences/workshops). He is serving as a reviewer for international journals and conferences including IEEE Transactions on Industrial Electronics, SN Applied Sciences, Energies and Complex & Intelligent Systems. He is also a guest editor for special issues in *Energies MDPI* and *Sustainability MDPI* journals.

Mahmoud A. Mossa, PhD, graduated and received his bachelor's and master's degrees in electrical engineering from the Faculty of Engineering, Minia University–Egypt in 2008 and 2013, respectively. From January 2010, he was working as an assistant lecturer at the electrical engineering department at the same university. In November 2014, he joined the electric drives laboratory (EDLAB) at University of Padova in Italy for his PhD research activities. In April 2018, he was awarded the PhD degree in electrical engineering from University of Padova. Since May 2018, he is working as an assistant professor at the electrical engineering department at Minia University in Egypt. He was a postdoctoral fellow at the department of industrial engineering at University of Padova in Italy for a 6-month started at October 2021. His research interests are focused on renewable energy systems, electric machine drives, power electronics, power management, and load frequency control. He published more than 60 papers (international journals, book chapters, and conferences/

workshops). He is an associate editor in *International Journal of Robotics and Control Systems*, *Journal of Robotics and Mechanical Engineering*, and *International Journal of Agriculture, Engineering Technology and Social Sciences*. He is a guest editor for launched special issues in the *Energies MDPI* and *Machines MDPI* journals. He served as a recognized reviewer for about 24 international journals and conferences.

Mariya Ouaissa, PhD, is currently a professor in cybersecurity and networks at the Faculty of Sciences Semlalia, Cadi Ayyad University, Marrakech, Morocco. She is a PhD graduate in 2019 in computer science and networks from the Laboratory of Modelisation of Mathematics and Computer Science from ENSAM-Moulay Ismail University, Meknes, Morocco. She is a networks and telecoms engineer and graduated in 2013 from National School of Applied Sciences Khouribga, Morocco. She is a co-founder and IT consultant at the IT Support and Consulting Center. She was working for the School of Technology of Meknes, Morocco, as a visiting professor from 2013 to 2021. She is member of the International Association of Engineers and the International Association of Online Engineering, and since 2021, she is an ACM professional member. She is an expert reviewer with Academic Exchange Information Centre (AEIC) and a brand ambassador with Bentham Science. She has served and continues to serve on technical program and organizer committees of several conferences and events and has organized many symposiums/workshops/conferences as a general chair also as a reviewer of numerous international journals. Dr. Ouaissa has made contributions in the fields of information security and privacy, Internet of Things security, and wireless and constrained networks security. Her main research topics are IoT, M2M, D2D, WSN, cellular networks, and vehicular networks. She has published over 50 papers (book chapters, international journals, and conferences/workshops), 13 edited books, and 8 special issues as guest editor.

Sanjeevikumar Padmanaban, PhD, (Member '12–Senior Member '15, IEEE) received a PhD degree in electrical engineering from the University of Bologna, Bologna, Italy, in 2012. He is a full professor in electrical power engineering with the Department of Electrical Engineering, Information Technology, and Cybernetics, University of South-Eastern Norway, Norway. S. Padmanaban has authored over 750+ scientific papers and received the Best Paper cum Most Excellence Research Paper Award from IET-SEISCON '13, IET-CEAT '16, IEEE-EECSI '19, IEEE-CENCON '19, and five best paper awards from ETAEERE '16 sponsored Lecture Notes in Electrical Engineering, Springer book. He is a fellow of the Institution of Engineers, India; the Institution of Electronics and Telecommunication Engineers, India; and the Institution of Engineering and Technology, United Kingdom. He received a lifetime achievement award from Marquis

Who's Who – USA 2017 for contributing to power electronics and renewable energy research. He is listed among the world's top two scientists (from 2019) by Stanford University, USA. He is an editor/associate editor/editorial board for refereed journals, in particular the *IEEE Systems Journal*, *IEEE Transaction on Industry Applications*, *IEEE ACCESS*, *IET Power Electronics*, *IET Electronics Letters*, and *Wiley-International Transactions on Electrical Energy Systems*; subject editorial board member – Energy Sources – *Energies Journal*, *MDPI*, and the subject editor for the *IET Renewable Power Generation*; *IET Generation, Transmission and Distribution*; and *FACETS Journal* (Canada).

Said Mahfoud, PhD, is currently a professor at the Higher School of Technology, Sultan Moulay Slimane University, Khenifra, Morocco. He graduated from ENSET of Mohammedia, Morocco, with a license in electrical engineering in 2012, and from ENSET of Rabat with a master's degree in 2015. He received his PhD in electrical engineering from Sidi Mohamed Ben Abdellaah University in 2022. From September 2013 to July 2023, he worked as an electrical engineering teacher at the Ministry of Education. His areas of interest are electrical engineering, power electronics, electronic engineering, wind energy, power systems, control systems engineering, design renewable energy, power electronics and drives electrical, power engineering, power systems modelling, reliability engineering, grid integration, photovoltaics systems, smart grid, wind turbines, machines control techniques drives, electric vehicles, machines fault diagnosis, artificial neural network, optimization algorithm, intelligent control machines. He is serving as a reviewer for international journals and conferences, including IEEE Transactions on Industrial Electronics, Access, journals on ISA transactions, and many journals in the MDPI.

Contributors

Benaissa Bellach
National School of Applied Sciences, University Mohammed I
Oujda, Morocco

Jyoti Bhattacharjee
University of Calcutta
Kolkata, India

Ouadiâ Chekira
Faculty of Science and Technology, USMBA
Fez, Morocco

Abdelhak Essounaini
Faculty of Sciences Ben M'Sik Sidi Othman, Hassan II University
Casablanca, Morocco

Amaresh Gantayat
Faculty of Engineering and Technology, Siksha Anusandhan
Bhubaneswar, India

Abdelfettah El-Ghajghaj
National School of Applied Sciences, Sidi Mohamed Ben Abdellah University
Fez, Morocco

Mohammed Ghammouri
National School of Applied Sciences, University Mohammed I
Oujda, Morocco

Mohamed Ali Hajjaji
Faculty of Sciences, University of Monastir
Monastir, Tunisia

Zakia Hammouch
National School of Applied Sciences, University Mohammed I
Oujda, Morocco

Mahdi Hermassi
Faculty of Sciences, University of Monastir
Monastir, Tunisia

Abdelilah Hilali
Faculty of Sciences, Moulay Ismail University
Meknes, Morocco

Najwa Jbira
Polydisciplinary Faculty of Khouribga, Sultan Moulay Slimane University of Beni Mellal
Beni-Mellal, Morocco

G. Jegadeeswari
AMET
Chennai, India

Hicham Karmouni
Nationals School of Applied Sciences of Marrakech
CadiiAyyad University
Marrakesh, Morocco

K. Kathiravan
Theni Kammavar Sangam College of Technology
Theni, Tamilnadu, India

B. Kirubadurai
Vel Tech Rangarajan Dr. Sagunthala R&D Institute of Science and Technology
Chennai, India

Youssef Kraiem
University Lille, Arts et Metiers Institute of Technology, Centrale Lille, Junia
Lille, France

Saber Krim
National Engineering School of Monastir, University of Monastir
Monastir, Tunisia

D. Lakshmi
AMET
Chennai, India

K. Lamnaouar
National School of Applied Sciences, University Mohammed I
Oujda, Morocco

Firyal Latrache
National School of Applied Sciences, University Mohammed I
Oujda, Morocco

Azeddine Loulijat
Faculty of Sciences and Technology, Hassan First University, FST of Settat
Settat, Morocco

Moncef El Marghichi
Faculty of Science, Abdelmalek Essaadi University
Tetouan, Morocco

Anita Mohanty
Silicon Institute of Technology, Bhubaneswar
Odisha, India

Subrat Kumar Mohanty
Electronic and Communication Engineering, Einstein Academy of Technology and Management
Odisha, India

Sudipta Mohanty
Odisha University of Technology and Research (OUTR)
Bhubaneswar, India

Ambarish G. Mohapatra
Silicon Institute of Technology, Bhubaneswar
Odisha, India

Sasmita Nayak
Mechanical Engineering, Government College of Engineering, Kalahandi
Odisha, India

Najib El Ouanjli
Faculty of Sciences and Technology, Hassan First University
Settat, Morocco

P. N. Rajnarayanan
Theni Kammavar Sangam College of Technology
Theni, Tamilnadu State, India

Subhasis Roy
University of Calcutta
Kolkata, India

Mhamed Sayyouri
National School of Applied Sciences, Sidi Mohamed Ben Abdellah University
Fez, Morocco

Wasswa Shafik
Dig Connectivity Research Laboratory (DCRLab)
Kampala, Uganda

Chapter 1

Introduction to Renewable Energy Technologies

Wasswa Shafik

1.1 INTRODUCTION

Renewable energy (RE) is produced from naturally replenishing sources with minimal negative environmental effects. Sunlight, wind, water, biomass, and geothermal heat are a few of these sources [1]. Technologies are used to harness and transform these renewable energy sources into valuable energy are called "renewable energy technologies." The term "similarity of renewable energy technologies (RETs)" refers to the many techniques, methods, and apparatuses employed in capturing and transforming energy from renewable sources into proper forms of power [2]. These technologies are essential for lowering greenhouse gas emissions, halting climate change, and changing the energy system to be more sustainable.

Solar photovoltaic (PV) systems use semiconductor materials to convert sunlight directly into power. Silicon-based photovoltaic cells absorb photons from sunlight to produce an electron flow that results in an electric current. It is possible to use this electricity immediately or store it in batteries for later use [3]. Solar PV systems are frequently placed on rooftops, solar farms, and other open areas. Mirrors or lenses are used in concentrated solar power (CSP) systems to focus sunlight onto a receiver, transforming the solar energy into heat. Steam is created, which powers a turbine attached to a generator to produce electricity [4]. Thermal storage devices are frequently used in CSP facilities to store excess heat, enabling electricity generation even when sunlight is unavailable, such as during cloudy conditions or nighttime.

Wind kinetic energy is captured by turbines and transformed into mechanical power. The turbine's turning blades power a generator, which generates electricity. In contrast to offshore wind turbines, which are situated in bodies of water like oceans or huge lakes with stronger and more reliable winds, onshore wind turbines are typically found on land [5]. Like this, hydropower creates electricity by harnessing the energy of moving water, such as that produced by rivers, dams, or ocean tides. Dams are used in traditional hydropower facilities to store water in reservoirs. Released water powers the turbines, which then turn generators to generate electricity [6]. While

DOI: 10.1201/9781003462460-1

pumped storage hydroelectric facilities store extra electricity by pumping water uphill and releasing it at peak demand, run-of-river hydropower systems generate electricity by diverting a portion of the river's flow.

Organic materials, including wood, agricultural waste, energy crops, and organic waste, are used to create biomass energy. Various techniques can transform biomass into heat, electricity, or biofuels [6]. Burning biomass creates heat that can be utilized for space heating or to create steam for electricity production. Biomass is gasified into a combustible gas that can be used to generate energy or heat buildings [1, 7]. Biogas, a renewable fuel, is created by the breakdown of organic waste during anaerobic digestion. Geothermal power plants go much further by using heat from deep below the Earth to create electricity. Hot water or steam is extracted from geothermal reservoirs via wells. To operate turbines attached to generators and generate energy, high-pressure steam or hot water is used [2, 3]. Depending on the temperature and make-up of the geothermal resource, there are several kinds of geothermal power plants, such as binary cycle plants, flash steam plants, and dry steam plants.

Globally, RE has acquired much momentum as nations realize they must switch to clean, sustainable energy sources, as demonstrated in Figure 1.1. The capacity of renewable energy has been rising quickly on a global scale [5]. By the end of 2020, the capacity of all renewable energy sources was expected to reach 2,799 gigawatts (GW), an increase of 10.3% from the year before. Globally, solar power has expanded dramatically. The growth of solar photovoltaic installations has been facilitated by declining costs, technological improvements, and supportive regulations [6]. Many nations, including China, the United States, and India, have greatly improved their solar power generation capacity, encouraging the building of both massive solar farms and rooftop installations [7].

In the global energy landscape, wind power has grown to be a significant factor. Numerous regions have witnessed significant expansion in onshore and offshore wind farms. The development of wind power capacity has been fueled by major investments made in wind energy projects by nations including China, the United States, Germany, and the United Kingdom [2, 4]. One of the biggest sources of renewable energy worldwide is still hydropower. Traditional hydroelectric facilities, such as huge dams and

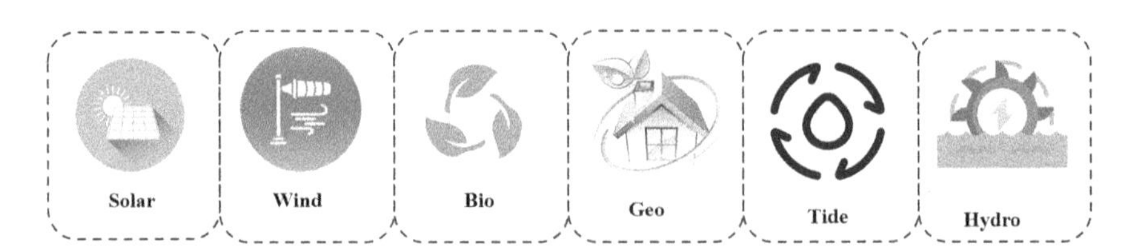

Figure 1.1 The main categories of RE sources.

run-of-river systems, nevertheless contribute significantly to the production of electricity [5]. However, attention is now being paid to less disruptive and more environmentally friendly types, like pumped storage and small-scale hydropower.

Bioenergy, which is produced from biomass sources, is a key component of the renewable energy portfolios of many nations – both the production of heat and power from biomass and the production of biofuels for transportation [8]. In many areas, there has been an increase in the use of agricultural wastes, forest biomass, and energy crops for the generation of bioenergy. Depending on the availability of geothermal resources, different countries utilize geothermal energy differently [9]. Leading nations that use the natural heat from the Earth's crust to create electricity include the United States, the Philippines, Indonesia, and Kenya.

Many nations have put supportive laws and incentives in place to promote the use of renewable energy. Renewable energy objectives, tax credits, subsidies, and feed-in tariffs have all been instrumental in encouraging the development and uptake of new renewable energy technology [10]. Global initiatives and accords, such as the Paris Agreement, have urged nations to use renewable energy sources and lower greenhouse gas emissions. International cooperation aims to exchange cutting-edge techniques, technological advancements, and financial assets to hasten the switch to renewable energy sources [11].

Even though RE has expanded significantly and in conversion, difficulties still exist. Grid integration, energy storage, and meeting the energy requirements of developing nations are a few of these [8, 11]. Nevertheless, it is anticipated that renewable energy will take center stage in the global energy mix because of continuous technological developments and a growing dedication to sustainability, helping to create a cleaner and more sustainable future. As shown later, RE is influencing the present energy landscape in a variety of ways.

Utilizing renewable energy assists in lowering greenhouse gas emissions, hence reducing the negative effects of climate change [11, 12]. Renewable energy technologies considerably reduce carbon footprints and help to achieve climate targets set in international agreements like the Paris Agreement by replacing fossil fuels, which are a major source of carbon dioxide emissions [13, 14]. Diversifying the energy mix with renewable energy lessens reliance on limited and imported fossil resources. By utilizing indigenous renewable resources, nations may increase their energy security and guarantee a more dependable and robust energy supply [1]. This reduces the dangers brought on by geopolitical unrest and changes in the price of fossil fuels.

Renewable energy sources have fewer negative environmental effects than fossil fuel-based power generation. They do not release dangerous pollutants like sulfur dioxide, nitrogen oxides, and particulates, which are linked to respiratory ailments, environmental deterioration, and air pollution [4].

Compared to traditional power plants, renewable energy technologies use less water, helping to preserve limited water resources. The market for renewable energy has become a substantial driver of economic expansion and employment development [5]. Infrastructure for renewable energy must be developed, installed, operated, and maintained by skilled workers, creating job possibilities. Additionally, financial support for renewable energy projects boosts economic growth by luring capital and fostering new ideas and technological developments [6].

Communities without access to conventional electrical grids may be able to receive contemporary energy services through RE. Off-grid renewable energy options, such as solar household systems and mini-grids, can strengthen rural and isolated communities by supplying clean, inexpensive electricity that supports social and economic development as well as health care and education [9]. The principles of sustainable development, which consider the environmental, social, and economic facets, are in line with RE. It encourages the development of healthier and cleaner environments, raises living standards, lessens poverty, and promotes inclusive growth [8]. Societies can achieve sustainable development objectives while guaranteeing intergenerational equity by embracing renewable energy.

The switch to RE stimulates technological development and innovation across a range of industries, as demonstrated in Figure 1.2. Improved renewable energy deployment is made possible by the development of more

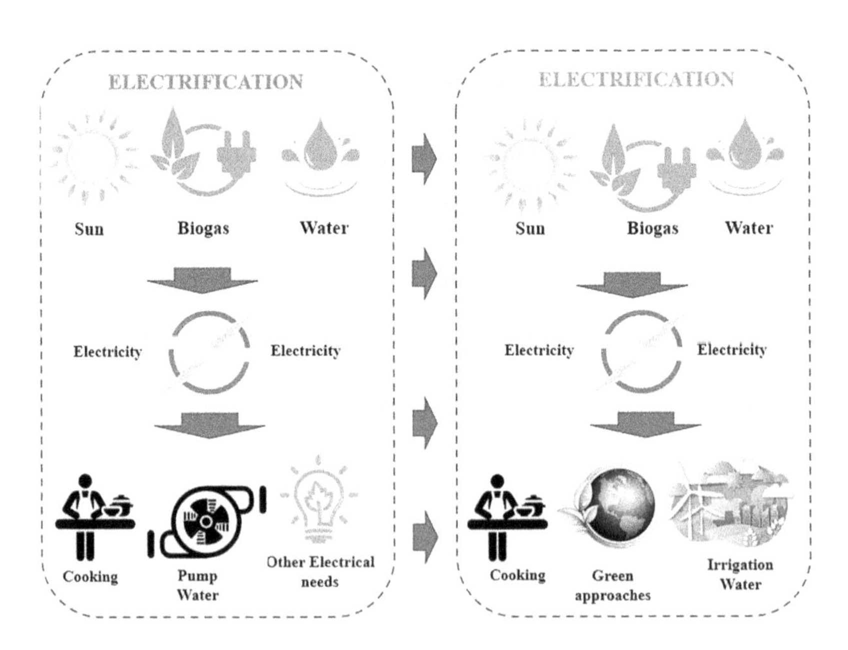

Figure 1.2 Electrification vs energization.

effective solar panels, sophisticated wind turbine designs, energy storage systems, and smart grid technologies [10]. These developments also encourage development in related industries, resulting in a more resilient and sustainable energy system. Therefore, given the current energy scenario, RE is of utmost significance [12, 13]. It combats climate change, improves energy security, safeguards the environment, generates jobs, boosts economic growth, increases access to energy, supports sustainable development, and advances technology [15]. Societies can attain a sustainable, low-carbon, and resilient energy future by accepting and scaling up RETs.

1.1.1 Chapter Contribution

This study provides the following contributions:

- The study presents a comprehensive overview of renewable energy technologies (RETs), emphasizing the significance of renewable energy in the current energy landscape.
- The study also serves as a foundation for further exploration and understanding of RETs entailing energy sources.
- The study systematically details the working principles of RETs, including solar energy (photovoltaic technology and concentrated solar power systems), wind energy (onshore and offshore turbines), hydropower (conventional, run-of-river, and pumped storage), biomass energy (different feedstocks and conversion technologies), and geothermal energy (different types of power plants).
- The study demonstrates several renewable energy technologies that are utilized in the renewable energy field.
- The study further discusses the applications, merits, and benefits of each renewable energy source.
- Based on the reviewed literature, the study identifies some future research directions that are more likely to shape the energy sector.
- It concludes by presenting some lessons learned from the renewable energy technologies discussed, highlighting the importance of transitioning to renewable energy and addressing prospects and challenges.

1.1.2 The Chapter Organization

Section 1.2 presents solar energy, encompassing an overview of solar energy, PV, and CSP technology. It also includes working principles, PV system types, applications, and benefits. Section 1.3 demonstrates wind energy, showing an overview of wind energy, onshore wind turbines, and types of onshore wind turbines entailing applications and benefits. Section 1.4 demonstrates the hydropower overview and conventional hydropower, run-of-river pumped storage hydropower, and types of conventional hydropower systems. Section 1.5 portrays the overview of biomass energy; types of biomass feedstocks, such as agricultural residues, forest biomass, and energy crops; and biomass conversion technologies, including anaerobic digestion, gasification, and combustion applications and benefits. Section 1.6 demonstrates the overview of geothermal energy, followed by geothermal power plants such as binary cycle, flash steam, and dry steam plants. Section 1.7 presents some selected RETs and demonstrates their application in the energy sector. Finally, Section 1.8 presents the future direction, lessons learned, and conclusion, entailing a summary of RETs discussed and the importance of transitioning to RE.

1.2 SOLAR AND ENERGY AND PHOTOVOLTAIC TECHNOLOGY

Within this section, we detail solar energy and photovoltaic technology.

1.2.1 Solar Energy

Solar energy is the use of sunlight to produce heat or power. It is one of the most accessible and abundant sources of RE [16]. Although solar energy has advanced significantly worldwide, it still faces difficulties such as grid integration, and the necessity for energy storage options. Solar energy is poised to play an increasingly significant role in the global energy mix, contributing to a more sustainable and resilient energy system, supportive policies, and the continued commitment to decarbonization [17].

1.2.1.1 Solar Photovoltaic Systems

Photovoltaic cells are used in solar PV systems to convert sunlight directly into electricity. Typically, silicon or other semiconductor materials are used to create these cells [18]. An electric current is produced when sunlight strikes PV cells because photons in the light energy knock electrons loose. Solar panels are made by combining several modules, which are joined together with various PV cells. Rooftops, solar farms, and even building materials can all be used to create PV systems [19].

1.2.1.2 Concentrated Solar Power Systems

CSP systems use mirrors or lenses to focus sunlight onto a receiver, transforming the solar energy into heat. The heat is then used to produce electricity using a variety of techniques [20]. Parabolic trough systems, tower systems, and dish/Stirling engine systems are examples of CSP technology. Large-scale power plants frequently employ CSP systems, which may also use thermal storage to provide electricity even when there is no sunshine [21].

1.2.1.3 Solar Heating and Cooling

Additionally, solar energy can be applied in heating and cooling systems. Water or other fluids heated by solar thermal systems can be utilized for room heating, water heating, or industrial activities [22]. Sun air conditioning systems replace the need for conventional cooling systems that use electricity by using solar energy to operate desiccant or absorption cooling processes [23]: the simplified scheme of the solar heating and cooling system, presenting the solar collectors, the absorption chiller, the cooling tower, the hot and cold storages, and the backup heat pump.

1.2.1.4 Solar Water Desalination

Desalination is another process that uses solar energy to turn salty or brackish water into freshwater [3]. Solar desalination systems use solar heat or solar PV systems to power the desalination process as a sustainable alternative for regions with scarce freshwater supplies.

1.2.1.5 Off-Grid Solar Systems

Off-grid solar systems, sometimes referred to as stand-alone solar systems, are intended to supply electricity in isolated locations without access to the electrical grid [24, 25]. The components of these systems include solar panels, energy-storing batteries, and inverters that transform DC power into AC power [4]. In rural areas, off-grid solar power systems are frequently utilized to power small-scale productive activities, appliances, and lighting.

1.2.1.6 Building-Integrated Photovoltaics

Solar PV modules are incorporated into building components like windows, roofs, and facades under the term building-integrated photovoltaics (BIPV) [9]. BIPV systems allow solar power generation to be seamlessly integrated into the built environment, allowing buildings to produce electricity without losing their visual appeal [5].

1.2.1.7 Solar Power Plants

Large-scale solar power facilities, sometimes known as solar farms, use many solar panels to produce electricity. These plants can be put on huge structures like solar canopies or floating platforms or mounted on the ground [10]. Solar power plants can provide energy to towns and industries while significantly contributing to grid-scale electricity generation. Numerous benefits come with solar energy, such as an almost endless supply, no emissions, and minimal operating expenses after the first investment in this metaverse era [26]. In addition to lowering carbon emissions and fostering energy independence, it is a crucial pillar in the transition to a clean and sustainable energy future [27]. Ongoing technology improvements, cost reductions, and supportive legislation fuel global solar energy adoption.

1.2.2 Photovoltaic Technology

Using photovoltaic cells, PV is a technique for producing electricity from sunshine. Through the photovoltaic effect, PV technology immediately transforms solar energy into electrical energy.

1.2.2.1 Photovoltaic Cells

The fundamental units of PV technology are photovoltaic cells, commonly known as solar cells. The most common semiconductor material used to make these cells is crystalline silicon [27]. When sunlight strikes a solar cell, photons from the light transfer their energy to the semiconductor material's electrons, causing them to get excited and resulting in an electric current [28].

1.2.2.2 PV Modules

PV modules, commonly referred to as solar panels, are constructed from interconnected photovoltaic cells that are enclosed. A single solar cell or several solar cells coupled in series or parallel can make up a PV module [29]. The modules are made to effectively convert sunlight into electricity and survive various environmental factors, including temperature variations and sun exposure [27].

1.2.2.3 System Components

PV systems have additional parts besides solar panels. Inverters are one of these components; they change the DC (direct current) electricity produced by the solar panels into AC (alternating current), which may be used in homes or businesses [28]. Other parts could be wiring, electrical switches, panel mounting frameworks, and meters to track energy output [15].

1.2.3 Types of PV Technologies

There are different types of PV technologies, including the following:

1.2.3.1 Crystalline Silicon (c-Si)

PV technology that uses crystalline silicon is the most popular. Monocrystalline silicon (mono-Si) and polycrystalline silicon are the categories it falls under (poly-Si). Polycrystalline cells are formed of numerous silicon crystals, whereas monocrystalline cells are made of a single silicon crystal structure [30]. Thin semiconductor layers are deposited on various substrates, such as glass, metal, or plastic, in thin-film photovoltaic technology. Amorphous silicon (a-Si), cadmium telluride (CdTe), and copper indium gallium selenide are examples of thin-film technologies (CIGS) [31]. Thin-film modules can be incorporated into building materials and are often lighter and more flexible. Figure 1.3 demonstrates the processes energy goes through from its sources through energy use and application.

1.2.3.2 Commercial and Industrial Residential

Residential structures frequently have rooftop solar panels installed to produce electricity for on-site use. Through net metering, extra energy can be returned to the grid or stored in batteries for later use. Commercial and industrial establishments use PV systems to offset electricity and lower energy costs [32]. Large-scale PV installations are also used in open areas and on commercial rooftops.

1.2.3.3 Utility-Scale Solar Energy Facilities

Utility-scale PV power plants produce electricity and are called solar farms or solar parks [31, 33]. These power plants generate electricity for the grid using many PV modules.

1.2.3.4 Off-Grid and Distant Areas

PV systems are employed in off-grid and remote locations with little or no electrical grid connectivity [34]. Stand-alone PV systems provide electricity for lighting, appliances, and other energy requirements with battery storage.

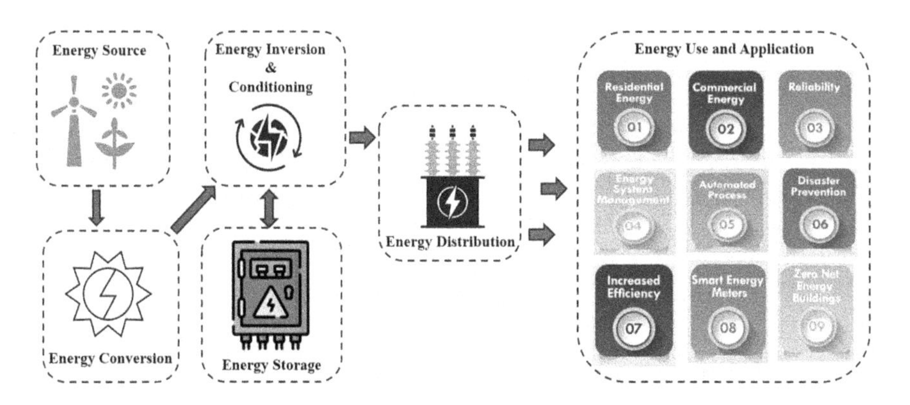

Figure 1.3 Energy Processes.

1.2.3.5 Renewable and Clean

Since solar energy is a renewable resource and PV technology generates power without releasing any greenhouse gases or other pollutants, it helps lower carbon emissions and improve air quality. PV systems can be scaled up or down depending on the energy needed [35]. It may be expanded with additional modules to improve capacity, making it adaptable for various applications [36]. Given that sunlight is abundant and has a long lifespan, PV systems have comparatively cheap operating and maintenance expenses once they are installed in uncertainty compensation, securing it against cyberattacks within smart energy networks [37–39].

1.2.3.6 Silent and Decentralized

Decentralized electricity generation is made possible by the quiet operation of PV systems, which may be deployed in various places, including rooftops, public spaces, and distant locales. Global use of solar PV technology has been tremendous [29, 40]. China, the United States, India, Japan, and Germany are leading the way in solar PV installations. Falling costs, governmental incentives, and environmental concerns have facilitated the global adoption of solar PV technology. Research in PV focuses on discovering novel materials and production techniques, increasing efficiency and lowering costs [15]. The acceptance of solar PV as a clean and sustainable source of electricity depends heavily on these developments.

1.2.4 Concentrated Solar Power Technology

CSP technology, sometimes referred to as solar thermal power, is a way to produce electricity by focusing sunlight through mirrors or lenses onto a receiver. The heat the focused sunlight creates is then utilized to generate energy using various methods.

1.2.4.1 Solar Concentration

Sunlight is focused onto a receiver using CSP technology, which uses mirrors or lenses. Solar energy may be converted into heat effectively because of the large increase in sunshine intensity brought on by this concentration [27]. The basic idea of CSP technology is to concentrate sunlight through lenses or mirrors and then use the focused heat to create electricity. CSP systems use mirrors or lenses to focus sunlight onto a particular region or receiver [1]. Mirrors are built to track the sun's movement throughout the day to maximize the quantity of sunlight focused on the receiver. The focal point of the mirrors or lenses serves as the receiver for the focused sunlight [2]. The receiver's purpose is to take in focused sunlight and turn it into heat.

1.2.4.1.1 Heat Transfer Fluid

A heat transfer fluid, like oil or molten salt, is typically cycled through the receiver in CSP systems. The heat transfer fluid absorbs thermal energy as concentrated sunlight strikes the receiver, heating it [5, 6].

1.2.4.1.2 Heat Storage (Optional)

A few CSP systems have thermal energy storage, which enables extra heat produced during the sun's peak hours to be stored for later use [9]. Molten salt, a heat transfer fluid that efficiently retains heat, can be used to store thermal energy.

1.2.4.1.3 Heat Exchange

After passing through a heat exchanger, the heated heat transfer fluid's thermal energy is converted to steam by being transferred to a working fluid, usually water. A turbine is the target of the heated working fluid, which is now steam. Turning the turbine blades with high-pressure steam transforms thermal energy into mechanical energy [10]. An electrical generator is attached to the rotating turbine and uses mechanical energy to produce electricity. The generated electricity can be fed into the electrical grid or used to power residences and commercial buildings [13].

1.2.4.1.4 Cooling and Condensation

Using a condenser, the steam that has been cooled and brought to a low pressure after passing through the turbine is converted to liquid form. The latent heat is released during condensation and is recoverable for use in other processes. Figure 1.4 illustrates the main RE diversity approaches that can be adopted to ensure proper energy utilization.

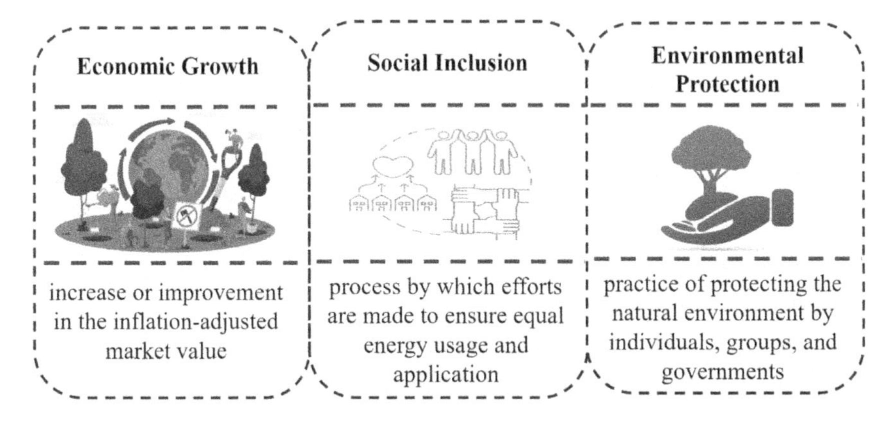

Figure 1.4 Top renewable energy diversity approaches.

1.2.4.1.5 *Water Treatment (If Necessary)*

Some CSP systems use condensed water from the steam that has been cleaned to eliminate pollutants before being repurposed to reduce water usage. Either on-site consumption or transmission to the electrical grid for distribution to customers are options for the generated electricity [41]. The CSP technology's basic tenets revolve around the effective concentration of sunlight, the conversion of solar energy into heat, and the use of heat to create steam and electricity. Incorporating thermal energy storage into CSP systems makes it possible for them to produce power even when there is not any sunlight, providing a dispatchable and dependable renewable energy option [42].

1.2.4.2 Types of CSP Systems

There are different types of CSP systems, including the following.

1.2.4.2.1 *Parabolic Trough Systems*

The most typical CSP technology is this one. Sunlight is focused onto a receiver tube that runs along the focal line using curved mirrors shaped like parabolic troughs [31]. The receiver tube's heat transfer fluid is heated as it passes through it, creating steam that powers a turbine attached to a generator to create electricity.

1.2.4.2.2 *Power Tower Systems*

Heliostats are a field of mirrors that follow the sun and reflect sunlight onto a central receiver suspended from a tower in power tower systems. A heat transfer fluid or molten salt in the receiver is heated by concentrated sunlight, and this heat is used to create steam and generate electricity [34].

1.2.4.2.3 *Stirling or Dish Engine Systems*

The focal point of this kind of CSP system is a receiver that receives sunlight through a big dish-shaped reflector [35]. A Stirling engine is housed inside the receiver, which uses the heat energy to power a generator to create electricity.

1.2.4.2.4 *Linear Fresnel Systems*

Sunlight is focused onto a linear receiver using linear Fresnel devices, which employ a sequence of flat mirrors. The receiver heats a working fluid, which produces steam and powers a turbine that generates energy [36].

1.2.4.2.5 Thermal Energy Storage

The capacity to use thermal energy storage is one benefit of CSP technology. Molten salt and other heat storage materials can be used to store extra heat produced during the height of the day [42]. When the sun is not shining, this thermal energy can be used to generate electricity, allowing for continuous power production [29].

1.2.4.2.6 Applications of CSP Technology

CSP technology can be used for large-scale power generation. It is especially well-suited to regions with high direct normal irradiance (DNI) and lots of available land [16]. Power from CSP plants can be dispatched, dependable, and integrated into existing power systems.

1.2.4.3 Advantages of CSP Technology

1.2.4.3.1 Dispatchable Power

CSP plants with thermal energy storage can produce dispatchable power because the heat they have stored can be used to produce electricity even when there isn't any sunshine, like at night or during cloudy conditions [32]. Integrating CSP technology into existing power networks can guarantee a steady and dependable electricity supply. Compared to solar systems, CSP systems offer the potential for greater efficiency because they can reach higher temperatures and use thermal energy storage.

1.2.4.3.2 Auxiliary Services

CSP facilities can offer grid ancillary services like frequency control and grid stability support.

Several nations, including Spain, the United States, Morocco, South Africa, and the United Arab Emirates, have implemented concentrated solar power technology. The growth of CSP projects has been fueled by the good solar resources and supporting legislation in these regions [43]. The main goals of ongoing CSP technology research and development are the advancement of thermal energy storage capacities, cost reduction, and efficiency [44]. These developments aim to strengthen CSP's position in transitioning to a clean energy future and make it more competitive with other renewable energy technologies [45].

The basic idea of CSP technology is to focus sunlight using mirrors or lenses and then harness the concentrated heat to produce energy. Here is a step-by-step explanation of the working principle. In CSP systems, sunlight is focused onto a certain region or receiver using mirrors or lenses [46]. To increase the amount of sunlight focused on the receiver, the mirrors are made to follow the sun's path throughout the day [47]. The receiver, which is situated at the focal point of the mirrors or lenses, receives concentrated sunlight. The receiver is made to take in the intense sunlight and turn it into heat.

Most CSP systems circulate a heat transfer fluid through the receiver, like oil or molten salt. The heat transfer fluid absorbs the thermal energy and warms up as the focused sunlight strikes the receiver [48]. Some CSP systems have thermal energy storage, which enables extra heat produced during the height of the solar day to be stored for later use. Thermo energy can be stored by using molten salt as the heat transfer fluid, which efficiently retains heat [49]. After passing through a heat exchanger, the heated heat transfer fluid's thermal energy is transferred to a working fluid, usually water, creating steam.

A turbine receives the heated working fluid, which has changed into steam. The turbine blades rotate because of the high-pressure steam, transforming thermal energy into mechanical energy [50]. The electrical generator, which is attached to the rotating turbine, transforms the mechanical energy into electricity. Electricity produced can be fed into the electrical grid or used to power buildings or homes [51]. The now-low-pressure, cooled steam is condensed back into liquid form by a condenser after passing past the turbine [52]. The latent heat is released during condensation and can be collected and applied elsewhere.

Some CSP systems treat condensed water from the steam to eliminate pollutants before reusing it in the system to reduce water use. Either the generated electricity is used locally, or it is sent to the electrical grid to be distributed to users [53]. The fundamental components of CSP technology are the efficient concentration of sunlight, the transformation of solar energy into heat, and the use of heat to create steam and electricity [54, 55]. By incorporating thermal energy storage, CSP systems may generate power even when there isn't any sunshine, providing a dispatchable and dependable renewable energy source.

1.3 WIND ENERGY

A renewable and sustainable energy source, geothermal energy is produced by the inherent heat of the Earth's interior. This heat is a result of the Earth's formation and the continuous radioactive decay of uranium, thorium, and potassium in the crust [54]. Various civilizations have used geothermal energy for heating and bathing for thousands of years, but in recent years it has gained importance as a cleaner alternative to fossil fuels for electricity generation and direct-use applications [28, 30].

1.3.1 Overview of Wind Energy

Onshore wind turbines are sophisticated devices built to harness and transform wind's kinetic energy into useful electricity. A detailed description of their operating concept is provided in the following.

1.3.1.1 Wind Capture

Aerodynamically designed rotor blades are constructed from lightweight, long-lasting materials like carbon fiber or fiberglass. They can absorb the most wind energy possible thanks to their curved shape and oblique angle of attack [56]. The blades revolve as the wind passes over them, producing a lift on one side and a low-pressure area on the other.

1.3.1.2 Gearbox and Transmission

The rotor blades rotate at a rather slow speed. The rotating speed must be increased for effective electricity generation. A gearbox is used to do this by increasing the speed and transmitting it to the generator [28]. In appreciation to the gearbox, the generator can run at the best speed for producing electricity.

1.3.1.3 Generator Operation

A crucial element that transforms mechanical spinning into electrical energy is the generator. Asynchronous or synchronous generators are used by most wind turbines (induction generators) [18]. In either scenario, the generator has wire coils that spin around in a magnetic field. Faraday's law of electromagnetic induction states that an electric current is induced in the coils when the wires pass through the magnetic field [19].

1.3.1.4 Electrical Conversion

Alternating current (AC), which corresponds to the frequency of the electrical grid, is initially used to create electricity. However, there are many frequencies and voltages at which wind turbines can generate power. Power electronics inside the nacelle, such as inverters, change the AC into the necessary frequency and voltage for grid compatibility [56].

1.3.1.5 Grid Connection

After being transformed, the electricity is sent into the grid via cables fed from the tower down. Wind turbines that are connected to the grid synchronize their frequency and voltage with the grid, making it possible to integrate and distribute the generated power to consumers without any interruptions [20].

1.3.1.6 Yaw and Pitch Control

The best performance is guaranteed by the control systems included in wind turbines. To maximize energy capture, the yaw system spins the entire nacelle to face the wind [1]. To control the turbine's rotational speed and power output, particularly during strong winds, blade pitch control modifies the angle of the rotor blades to prevent damage. The combination of the presented overview could result in wind energy from a wind turbine and different components.

1.3.2 Types of Onshore Wind Turbines

1.3.2.1 Horizontal Axis Wind Turbines (HAWTs)

The most prevalent kind of onshore wind turbines are HAWTs. Their rotor blades spin parallel to the ground and feature a horizontal main shaft. Based on where the rotor and nacelle are located, HAWTs are further divided into several categories, including upwind turbines. Upwind turbines have a rotor that faces the wind and a nacelle that is upwind of the tower and houses the gearbox and generator [3]. This design is more widely used because of its greater effectiveness and better wind capture. With downwind turbines, the rotor of these turbines is positioned to the side of the tower. In this design, the nacelle and blades are oriented away from the wind [8]. Although this design makes the mechanical structure simpler and less susceptible to damage from unforeseen gusts, it is typically less effective due to turbulence brought on by the tower [7].

1.3.2.2 Vertical Axis Wind Turbines (VAWTs)

VAWTs have a vertical main shaft around which their rotor blades revolve. Although less prevalent than HAWTs, they have advantages like the capacity to capture wind from all directions. VAWTs can be broken down into various subtypes [11].

1.3.2.2.1 Darrieus Turbines

Darrieus turbines have airfoil-shaped blades that are vertically orientated and resemble an eggbeater. The turbine spins because the rotating blades produce lift like an airplane wing. Darrieus turbines are ideal for both low and high wind speeds and are self-starting. They are frequently applied in little-scale projects [13].

1.3.2.2.2 Savonius Turbines

These simpler-looking turbines have bent blades that generate drag as they rotate. Despite being less effective, this design has the benefit of being self-sufficient. Due to their simplicity, Savonius turbines are frequently employed in urban settings and for educational reasons [29].

1.3.2.3 Hybrid Turbines

Wind turbines that are hybrids combine advantages of both VAWTs and HAWTs. The objective is to capture wind energy from various wind directions and speeds, hence improving total efficiency [4]. Due to growing mechanical complexity and difficulties with performance optimization, hybrid designs are less prevalent [5]. Several variables, such as wind speed, available space, efficiency objectives, and financial constraints, influence the wind turbine type. Due to their greater efficiency and extensive deployment, HAWTs are more common [10, 11]; they work well in places where the wind blows in the same general direction.

On the other hand, VAWTs can be effective in urban settings and places where the direction of the wind is varied [6]. Wind turbine design innovation is still being pursued to boost productivity, save costs, and address particular noise, aesthetics, and intermittent operation issues. Regardless of the kind, onshore wind turbines are essential for capturing clean, renewable energy and significantly impact the global shift to sustainable energy sources [15].

1.3.3 Wind Turbine Applications

1.3.3.1 Electricity Generation

By using the wind's kinetic energy to generate clean, renewable energy, onshore wind energy substantially contributes to electricity production. The mechanical energy produced by revolving blades is converted into electrical energy by wind turbines outfitted with aerodynamically constructed rotor blades and effective generators [57]. Wind provides lift and rotational motion as it moves across the blades, powering the generator's rotor and generating energy. Reducing greenhouse gas emissions, air pollution, and dependence on depletable fossil fuels are just a few advantages of this application [58]. Onshore wind farms can be carefully positioned in places with good wind conditions, offering a reliable source of energy that helps create a more resilient and sustainable energy mix [33]. However, due to the sporadic nature of wind, other energy sources or energy storage technologies are needed to provide a consistent and dependable power supply.

1.3.3.2 Community Power

Onshore wind farms with a community focus enable locals and cooperatives to participate in energy production and ownership. In this application, communities jointly own and operate wind turbines, encouraging a sense of shared responsibility for generating renewable energy [32]. These programs frequently allow localities to profit directly from the cash generated, which can then be used to fund infrastructure upgrades, local development initiatives, or educational programs [29, 59]. A more decentralized and democratic approach to energy generation is promoted by community involvement in wind projects, in addition to the economic benefits. Establishing community projects, however, necessitates navigating difficult financial issues, maintaining effective stakeholder engagement, and addressing a range of community interests [32].

1.3.3.3 Grid Support and Stabilization

The potential of onshore wind energy to inject electricity during times of high demand and support overall system balancing adds to grid stability. By adjusting their power output in response to system circumstances, wind farms with smart grid technology can assist in controlling demand fluctuations and prevent interruptions [60]. Wind power eliminates the need for traditional power plants, which are frequently less efficient and release more pollutants, by producing electricity at times of peak demand. Furthermore, while grid integration solutions like energy storage systems and demand response mechanisms can reduce this fluctuation and provide a consistent power supply, wind energy is changeable due to changes in wind speed [44]. Cooperation between grid and wind farm operators is essential to integrate wind energy into the greater energy ecosystem.

1.3.3.4 Water Pumping and Desalination

Water-related applications, such as powering water pumps for irrigation, cattle watering, and community water supply, demonstrate the versatility of onshore wind energy. Wind-powered water pumping is a sustainable solution for distant areas or regions without access to traditional power sources [46, 61]. These systems deliver a dependable water supply while reducing their negative effects on the environment by using wind energy. Wind energy can also power desalination procedures, which turn saltwater into freshwater. This application is especially helpful in coastal areas with limited access to clean water [46, 47]. However, the changeable wind energy output must be coordinated with the capacity for water pumping or desalination, and maintenance issues in remote areas must also be considered.

1.4 RENEWABLE ENERGY TECHNOLOGIES

RETs encompass a range of methods and systems that harness naturally occurring resources to generate clean and sustainable energy. These technologies offer alternatives to fossil fuels and contribute to mitigating climate change and reducing environmental impacts.

1.4.1 Solar Photovoltaic Systems

Solar PV systems turn sunlight directly into power by utilizing the photovoltaic effect. Sun panels are made up of many silicon-based semiconductor solar cells. An electric current is created when photons from the sunlight strike the solar cells, releasing electrons from their atoms [51].

Inverters then transform the DC into AC for usage in residences, commercial buildings, and the grid. Rooftop installations, solar farms, and solar-powered gadgets are a few examples of the various solar PV system configurations [55].

1.4.2 Solar Thermal Systems

Solar thermal systems use solar energy to heat fluids inside solar collectors, usually water or heat transfer fluids. The steam these heated fluids produce is subsequently utilized to power turbines attached to generators, generating electricity [22]. Solar thermal technology can also supply direct heating for domestic, commercial, and industrial uses such as space heating, water heating, and industrial operations.

1.4.3 Wind Power

Utilizing the kinetic energy of moving air masses, wind power produces electricity. Rotor blades that are coupled to a central hub make up wind turbines. The blades revolve due to the lift produced by the wind as it passes over them. A shaft that is attached to a generator is turned by this rotational energy, which generates electricity [48, 62]. Onshore or offshore wind farms are also possible, with offshore farms frequently producing higher wind speeds and capacity factors.

1.4.4 Hydropower

The gravitational potential energy of bodies of water is the basis for hydropower. River water is diverted to turbines by dams or water intakes. The turbines rotate due to the potential energy released by the water as it passes past them. Generators are powered by rotational energy to generate electricity [50]. The two types of hydropower systems are reservoir-based systems, which use water stored in dams for regulated energy generation, and run-of-the-river systems, which use natural river flow without extensive storage [63].

1.4.5 Geothermal Energy

The heat produced by the decay of radioactive isotopes and the planet's natural heat generates geothermal energy. Wells transport hot water or steam from underground geothermal reserves to the surface [46]. This steam powers turbines to produce energy. Through geothermal heat pumps, geothermal energy also provides direct warmth for structures and industrial processes; four RE sustainabilities are demonstrated in Figure 1.5.

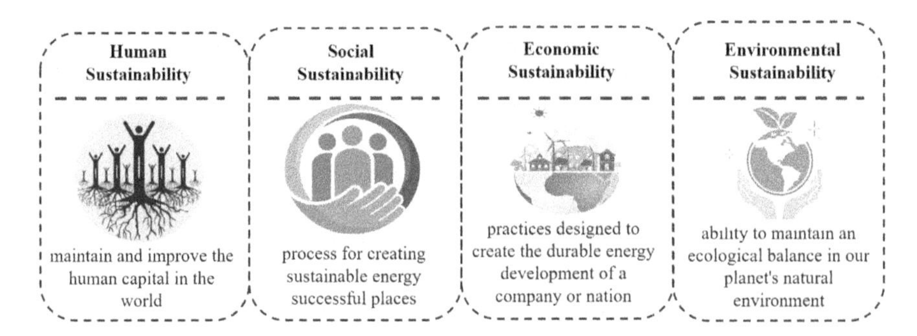

Figure 1.5 Top RET sustainable aspects.

1.4.6 Biomass Energy

Utilizing organic materials like wood, crop byproducts, and agricultural waste to produce heat, electricity, or biofuels is known as biomass energy. Direct combustion of biomass in power plants is also an option, as is fermentation and anaerobic digestion for producing biogas or biofuels from it [42, 64]. Utilizing organic material sustainably and lowering the carbon footprint associated with the disposal of organic waste are both possible with biomass energy.

1.4.7 Ocean Energy

Technologies that capture the energy from tides, waves, and ocean currents are referred to as ocean energy. Tidal energy can be harnessed by installing tidal barrages, which entail damming tidal estuaries to provide regulated flow, or tidal stream generators, which are submerged windmills positioned in locations with high tidal currents [32]. Wave energy devices capture the energy from the up-and-down motion of waves. Underwater turbines are used in ocean current energy systems to produce power.

1.4.8 Hybrid Systems

Systems that combine two or more renewable energy sources maximize energy production and dependability. A wind-solar hybrid system, for instance, can generate electricity when both wind and solar resources are present [1]. To store excess energy for use during times of low generation, hybrid systems frequently integrate energy storage components.

1.4.9 Energy Storage

Technologies for energy storage are essential for coping with the intermittent nature of some renewable energy sources. Batteries store extra energy produced by renewable sources during periods of high generation and release it during periods of low generation or high demand [3]. Pumped hydro storage includes pumping water uphill when there is little demand for energy and releasing it through turbines when there is [6]. Extra heat or cold is stored thermally for later use.

1.4.10 Smart Grids and Demand Response

Digital communication and cutting-edge control systems are incorporated into the conventional electrical grid infrastructure by smart grids. They make it possible to regulate and analyze energy flows in real-time, which improves the integration of renewable energy sources and boosts load-balancing effectiveness [15]. Demand response programs encourage customers to modify their energy use in reaction to current power prices and system circumstances, relieving pressure during high-demand periods.

1.5 FUTURE RESEARCH DIRECTION, LESSONS LEARNED, AND THE CONCLUSION

Within this section, we present the future research directions, some lessons learned from the study, and a conclusion in RETs that encompass a wide range of areas, as demonstrated next.

1.5.1 Future Research Direction

Future research in RETs will focus on various topics that will advance and enhance many facets of the production, integration, and storage of renewable energy. We detail the eight most important future research directions in this subsection.

1.5.1.1 Energy Storage Technologies

Energy storage research aims to create cutting-edge storage systems that can effectively store extra renewable energy for later use. It also covers upcoming battery technologies like hydrogen storage, compressed air energy storage, and thermal energy storage, as well as improvements in battery technologies like next-generation lithium-ion batteries, solid-state batteries, flow batteries, and others [28, 65]. The intermittent character of renewable energy sources must be overcome to provide a steady and dependable energy supply, which calls for improved energy storage technologies [66].

1.5.1.2 Grid Integration and Energy Management

Research is required to create advanced grid integration strategies and energy management systems as the use of renewable energy sources spreads. To balance supply and demand and increase grid stability, this calls for integrating renewable energy into current power networks, creating smart grid technologies, improving demand-side management, and implementing advanced control systems [41].

1.5.1.3 Advanced Solar Technologies

Future solar technology research will concentrate on increasing the effectiveness, performance, and cost-effectiveness of solar energy conversion. Achieving higher efficiency and reduced manufacturing costs includes the development of next-generation PV technologies such as perovskite solar cells, tandem solar cells, and innovative materials [34]. Research also focuses on cutting-edge solar thermal technologies, such as novel materials for enhanced heat absorption and storage and sophisticated CSP systems.

1.5.1.4 Wind Energy

Future wind energy research intends to increase the effectiveness and dependability of wind turbines, lower maintenance costs, and investigate new sites for the development of wind farms, such as offshore and high-altitude wind power [31]. This entails enhancing the design and composition of turbines, creating sophisticated control systems for improved grid integration, and enhancing the layout of wind farms to maximize energy output [52].

1.5.1.5 Bioenergy and Biofuels

Bioenergy research focuses on advanced biofuels and biogas technologies that can replace fossil fuels in various industries, including transportation and manufacturing. Improvements in biomass conversion technologies, such as thermochemical and biochemical procedures, as well as the investigation of viable feedstock sources and effective biomass production and harvesting techniques, are all included in this [67].

1.5.1.6 Electrification of Transportation

One important area of research to lower greenhouse gas emissions and reliance on fossil fuels is the electrification of transportation. Future research will concentrate on enhancing EVs' effectiveness, range, and charging infrastructure; creating cutting-edge batteries; investigating alternative fuel cell technologies; and implementing smart charging and vehicle-to-grid integration for the best possible energy management [61].

1.5.1.7 Power-to-X Technologies

Power-to-X solutions include transforming extra renewable energy into useful forms like heat, heat, or hydrogen. Future research in this field aims to create efficient and affordable electrolysis processes that produce green hydrogen, advance catalytic techniques that transform CO_2 into useful chemicals or synthetic fuels and investigate the integration of power-to-X technologies with renewable energy systems to create a carbon-neutral energy cycle [56].

1.5.1.8 Sustainable Materials and Manufacturing

Reducing the environmental effect of energy production and enhancing the lifecycle sustainability of renewable energy systems are the goals of research in sustainable materials and manufacturing for renewable energy technologies [20]. This entails investigating environmentally friendly materials for use in solar panels, wind turbine parts, and energy storage devices and creating environmentally friendly manufacturing techniques that use less energy, produce less waste, and emit fewer carbon emissions [18]. The advancement of renewable energy technologies, assuring their long-term viability, and hastening the transition to a sustainable and low-carbon energy future all depend on these research directions [68]. The lessons discovered here are presented in the subsection that follows.

1.5.2 Lessons Learned from the Chapter

Lessons learned from the development and deployment of RETs offer valuable insights into their implementation, challenges faced, and areas for improvement.

1.5.2.1 Policy and Regulatory Support

The significance of enabling policy and regulatory frameworks is one of the important lessons discovered. Through programs like feed-in tariffs, tax breaks, renewable portfolio standards, and net metering, governments are vital in encouraging renewable energy use [53, 69]. Long-term market signals are provided by transparent and stable policies, which also lower investment risks and promote sector expansion.

1.5.2.2 Technological Advancements

Technology has advanced significantly in the field of renewable energy throughout time. Lessons learned emphasize the value of ongoing research and development initiatives to raise the effectiveness, lower the cost, and increase the performance and dependability of renewable energy systems [48]. Renewable energy's competitiveness in the energy sector is fueled by ongoing innovation.

1.5.2.3 Grid Integration and Energy Storage

Due to its erratic nature, integrating renewable energy into current systems is difficult. The need for effective grid planning, grid coding, and grid strengthening to accommodate larger shares of renewable energy is emphasized by lessons learned [51, 70]. Energy storage systems are also essential for controlling fluctuations and providing a dependable and consistent power supply, especially during times when the production of renewable energy is low [40].

1.5.2.4 Local Community Engagement

The importance of incorporating local populations in renewable energy projects is highlighted by lessons learned. Early community involvement in planning and development can address issues, foster support, ensure local involvement, and provide benefits, resulting in a smoother project implementation and more social acceptance [20].

1.5.2.5 Public Awareness and Education

The extensive adoption of renewable energy solutions depends on public understanding and education. Lessons learned stress the value of spreading information, highlighting the advantages of renewable energy, and encouraging public involvement through awareness campaigns, educational efforts, and community outreach projects [28].

1.5.2.6 Collaborative Partnerships

For renewable energy projects to be implemented successfully, cooperation between numerous stakeholders, including governments, businesses, academics, and communities, is essential [50]. Lessons learned highlight the importance of creating partnerships to take advantage of experience, share knowledge, mobilize resources, and tackle problems as a group. Collaboration can hasten the shift to an economy based on renewable energy [32].

1.5.2.7 Economic Viability and Cost Reduction

As a result of tremendous cost reductions in recent years, renewable energy technologies are becoming more and more competitive with traditional energy sources. Lessons learned emphasize scale antifinance of accelerating the implementation of renewable energy sources, fostering economies of scale, and utilizing technology developments to drive down costs further [45, 71]. The long-term economic viability of renewable energy sources is guaranteed by ongoing attempts to increase cost-effectiveness.

1.5.2.8 Environmental Sustainability and Social Benefits

RETs positively impact the environment since they lower greenhouse gas emissions and enhance air quality. Lessons acquired underline the necessity of maximizing these environmental advantages while considering social factors [54]. For renewable energy projects to be built sustainably, biodiversity must be protected, any environmental repercussions must be addressed, and local populations must receive social and economic benefits. Policymakers, financiers, developers, and communities involved in the development and execution of renewable energy projects can benefit greatly from these lessons gained [21]. These insights can help the renewable energy industry expand, support efforts to reduce carbon emissions globally, and pave the way for a sustainable energy future.

1.5.3 Conclusion

This research introduced RETs by examining their definitions of various RETs and examples. It covered the significance of renewable energy in the existing energy system, emphasizing its contribution to reducing global warming, boosting energy security, and encouraging sustainable development. The chapter then went into detail about solar energy, outlining its fundamentals, the state of the world today, and major technological developments in the sector. The renewable energy industry has advanced significantly in recent years thanks to legislation encouraging it, technological improvements, and rising public awareness. Intermittency, grid integration, and cost competitiveness are among the ongoing difficulties. To progress renewable energy technology, policymakers, researchers, and industry stakeholders must work together and focus research in vital areas. Research efforts should concentrate on energy storage technologies, grid integration, and energy management, advancements in solar, wind, and bioenergy technologies, electrification of transportation, power-to-X technologies, and sustainable materials and manufacturing to drive future renewable energy development. We can overcome current obstacles, improve the functionality and effectiveness of renewable energy technology, and hasten the world's transition to a cleaner, more sustainable energy system by addressing these research directions. RETs have enormous promise for supplying the world's expanding energy demands while reducing the negative effects of climate change. To fully utilize the advantages of renewable energy and open the door to a sustainable and carbon-neutral future, investing in ongoing research and innovation, supporting policies, and stakeholder collaboration is imperative.

This chapter highlights the significance of incorporating renewable energy into the current energy infrastructure. It emphasizes the role of renewable energy in mitigating global warming, bolstering energy resilience, and fostering sustainable progress. An in-depth analysis of solar energy was offered, encompassing its basic principles, present worldwide situation, and significant technological breakthroughs in the field. This chapter recognizes the substantial gains made in the renewable energy sector, which can be ascribed to favorable legislation, technological innovations, and heightened public consciousness. The renewable energy sector still encounters obstacles such as intermittency, grid integration issues, and cost competitiveness. The significance of cooperation among policymakers, researchers, and industry stakeholders is emphasized. There is support of focused research initiatives in critical domains encompassing energy storage technologies, integration of power grids, breakthroughs in diverse renewable energy technologies, electrification of transportation, power-to-X technologies, and sustainable materials and manufacturing. In the future, tackling the renewable energy and green technology study areas can overcome the existing challenges encountered by renewable energy technologies. Sustained investment in ongoing research and innovation, the adoption of favorable legislation, and cooperation among stakeholders to fully exploit the benefits of renewable energy is needed to shift toward a cleaner, more sustainable, and carbon-neutral energy system.

REFERENCES

[1] Muwanga, R., Philemon Mwiru, D., & Watundu, S. (2023). Influence of social-cultural practices on the adoption of Renewable Energy Technologies (RETs) in Uganda. *Renewable Energy Focus*, 45. https://doi.org/10.1016/j.ref.2023.04.004

[2] Solangi, Y. A., Longsheng, C., & Shah, S. A. A. (2021). Assessing and overcoming the renewable energy barriers for sustainable development in Pakistan: An integrated AHP and fuzzy TOPSIS approach. *Renewable Energy*, 173. https://doi.org/10.1016/j.renene.2021.03.141

[3] Maradin, D., Cerović, L., & Mjeda, T. (2017). Economic effects of renewable energy technologies. *Naše Gospodarstvo/Our Economy*, 63(2). https://doi.org/10.1515/ngoe-2017-0012

[4] Liang, Y., Ju, Y., Martínez, L., Dong, P., & Wang, A. (2022). A multi-granular linguistic distribution-based group decision making method for renewable energy technology selection. *Applied Soft Computing*, 116. https://doi.org/10.1016/j.asoc.2021.108379

[5] Raza, M. Y., Wasim, M., & Sarwar, M. S. (2020). Development of renewable energy technologies in rural areas of Pakistan. In *Energy Sources, Part A: Recovery, Utilization and Environmental Effects* (Vol. 42, Issue 6). https://doi.org/10.1080/15567036.2019.1588428

[6] Patmal, M. H., & Shirani, H. (2021). Public awareness and their attitudes toward adopting renewable energy technologies in Afghanistan. *International Journal of Innovative Research and Scientific Studies*, 4(2). https://doi.org/10.53894/ijirss.v4i2.61

[7] Shafik, W. (2024). Blockchain-based internet of things (B-IoT): Challenges, solutions, opportunities, open research questions, and future trends.

Blockchain-based Internet of Things, 1, 35–58. Taylor & Francis. https://doi.org/10.1201/9781003407096-3
[8] de Melo, C. A., da Silva, M. P., & da Silva Benedito, R. (2021). Renewable energy technologies: Patent counts and considerations for energy and climate policy in Brazil. *Climate and Development*, 13(7). https://doi.org/10.1080/17565529.2020.1848778
[9] Ribeiro, F., Ferreira, P., Araújo, M., & Braga, A. C. (2014). Public opinion on renewable energy technologies in Portugal. *Energy*, 69. https://doi.org/10.1016/j.energy.2013.10.074
[10] Alam Hossain Mondal, M., Kamp, L. M., & Pachova, N. I. (2010). Drivers, barriers, and strategies for implementation of renewable energy technologies in rural areas in Bangladesh-An innovation system analysis. *Energy Policy*, 38(8). https://doi.org/10.1016/j.enpol.2010.04.018
[11] Rani, P., Mishra, A. R., Pardasani, K. R., Mardani, A., Liao, H., & Streimikiene, D. (2019). A novel VIKOR approach based on entropy and divergence measures of Pythagorean fuzzy sets to evaluate renewable energy technologies in India. *Journal of Cleaner Production*, 238. https://doi.org/10.1016/j.jclepro.2019.117936
[12] Aikhuele, D. O., Ighravwe, D. E., & Akinyele, D. (2019). Evaluation of renewable energy technology based on reliability attributes using hybrid fuzzy dynamic decision-making model. *Technology and Economics of Smart Grids and Sustainable Energy*, 4(1). https://doi.org/10.1007/s40866-019-0072-2
[13] Retnanestri, M., & Outhred, H. (2013). Acculturation of renewable energy technology into remote communities: Lessons from Dobrov, Bourdieu, and Rogers and an Indonesian case study. *Energy, Sustainability and Society*, 3(1). https://doi.org/10.1186/2192-0567-3-9
[14] Wassie, Y. T., & Adaramola, M. S. (2019). Potential environmental impacts of small-scale renewable energy technologies in East Africa: A systematic review of the evidence. In *Renewable and Sustainable Energy Reviews* (Vol. 111). https://doi.org/10.1016/j.rser.2019.05.037
[15] Nchofoung, T. N., Fotio, H. K., & Miamo, C. W. (2023). Green taxation and renewable energy technologies adoption: A global evidence. *Renewable Energy Focus*, 44. https://doi.org/10.1016/j.ref.2023.01.010
[16] Wang, Y., Shafik, W., Seong, J. T., Al Mutairi, A., Mustafa, M. S., & Mouhamed, M. R. (2023). Service delay and optimization of the energy efficiency of a system in fog-enabled smart cities. *Alexandria Engineering Journal*, 84, 112–125. https://doi.org/10.1016/j.aej.2023.10.034
[17] Shafik, W. (2023). IoT-based energy harvesting and future research trends in wireless sensor networks. *Handbook of Research on Network-Enabled IoT Applications for Smart City Services*, 282–306. https://doi.org/10.4018/979-8-3693-0744-1.ch016
[18] Levenda, A. M., Behrsin, I., & Disano, F. (2021). Renewable energy for whom? A global systematic review of the environmental justice implications of renewable energy technologies. *Energy Research and Social Science*, 71. https://doi.org/10.1016/j.erss.2020.101837
[19] Persoon, P. G. J., Bekkers, R. N. A., & Alkemade, F. (2022). The knowledge mobility of Renewable Energy Technology. *Energy Policy*, 161. https://doi.org/10.1016/j.enpol.2021.112670

[20] Siksnelyte-Butkiene, I., Zavadskas, E. K., & Streimikiene, D. (2020). Multi-criteria decision-making (MCDM) for the assessment of renewable energy technologies in a household: A review. *Energies*, 13(5). https://doi.org/10.3390/en13051164
[21] Qazi, A., Hussain, F., Rahim, N. A. B. D., Hardaker, G., Alghazzawi, D., Shaban, K., & Haruna, K. (2019). Towards sustainable energy: A systematic review of renewable energy sources, technologies, and public opinions. *IEEE Access*, 7. https://doi.org/10.1109/ACCESS.2019.2906402
[22] Kardooni, R., Yusoff, S. B., & Kari, F. B. (2016). Renewable energy technology acceptance in Peninsular Malaysia. *Energy Policy*, 88. https://doi.org/10.1016/j.enpol.2015.10.005
[23] Shafik, W. (2024). Toward a more ethical future of artificial intelligence and data science. In *The Ethical Frontier of AI and Data Analysis* (pp. 362–388). IGI Global. https://doi.org/10.4018/979-8-3693-2964-1.ch022
[24] Ayadi, O., Mauro, A., Aprile, M., & Motta, M. (2012). Performance assessment for solar heating and cooling system for office building in Italy. *Energy Procedia*, 30, 490–494. https://doi.org/10.1016/j.egypro.2012.11.058
[25] AlMallahi, M. N., Asaad, S. M., Inayat, A., Harby, K., & Elgendi, M. (2023). Analysis of solar-powered adsorption desalination systems: Current research trends, developments, and future perspectives. *International Journal of Thermofluids*, 100457. https://doi.org/10.1016/j.ijft.2023.100457
[26] Shafik, W. (2024). Predicting future cybercrime trends in the metaverse era. In *Forecasting Cyber Crimes in the Age of the Metaverse* (pp. 78–113). IGI Global. https://doi.org/10.4018/979-8-3693-0220-0.ch005
[27] Akinwale, Y. O., & Adepoju, A. O. (2019). Factors influencing willingness to adopt renewable energy technologies among micro and small enterprises in Lagos state Nigeria. *International Journal of Sustainable Energy Planning and Management*, 19. https://doi.org/10.5278/ijsepm.2019.19.7
[28] Shafik, W. (2023). Making cities smarter: IoT and SDN applications, challenges, and future trends. In *Opportunities and Challenges of Industrial IoT in 5G and 6G Networks* (pp. 73–94). IGI Global. https://doi.org/10.4018/978-1-7998-9266-3.ch004
[29] Lee, C. C., Yuan, Z., Lee, C. C., & Chang, Y. F. (2022). The impact of renewable energy technology innovation on energy poverty: Does climate risk matter? *Energy Economics*, 116. https://doi.org/10.1016/j.eneco.2022.106427
[30] Masukujjaman, M., Alam, S. S., Siwar, C., & Halim, S. A. (2021). Purchase intention of renewable energy technology in rural areas in Bangladesh: Empirical evidence. *Renewable Energy*, 170. https://doi.org/10.1016/j.renene.2021.01.125
[31] Hailemariam, A., Ivanovski, K., & Dzhumashev, R. (2022). Does R&D investment in renewable energy technologies reduce greenhouse gas emissions? *Applied Energy*, 327. https://doi.org/10.1016/j.apenergy.2022.120056
[32] Chel, A., & Kaushik, G. (2018). Renewable energy technologies for sustainable development of energy efficient building. *Alexandria Engineering Journal*, 57(2). https://doi.org/10.1016/j.aej.2017.02.027
[33] Feng, C., Wang, Y., Kang, R., Zhou, L., Bai, C., & Yan, Z. (2021). Characteristics and driving factors of spatial association network of China's renewable energy technology innovation. *Frontiers in Energy Research*, 9. https://doi.org/10.3389/fenrg.2021.686985

[34] Xin, L., Sun, H., Xia, X., Wang, H., Xiao, H., & Yan, X. (2022). How does renewable energy technology innovation affect manufacturing carbon intensity in China? *Environmental Science and Pollution Research*, 29(39). https://doi.org/10.1007/s11356-022-20012-8
[35] Yan, Z., Zou, B., Du, K., & Li, K. (2020). Do renewable energy technology innovations promote China's green productivity growth? Fresh evidence from partially linear functional-coefficient models. *Energy Economics*, 90. https://doi.org/10.1016/j.eneco.2020.104842
[36] Uddin, R., Khan, H. R., Arfeen, A., Shirazi, M. A., Rashid, A., & Khan, U. S. (2021). Energy storage for energy security and reliability through renewable energy technologies: A new paradigm for energy policies in Turkey and Pakistan. *Sustainability*, 13(5). https://doi.org/10.3390/su13052823
[37] Nasab, M. A., Zand, M., Dashtaki, A. A., Nasab, M. A., Padmanaban, S., & Blaabjerg, F. (2023). Uncertainty compensation with coordinated control of EVs and DER systems in smart grids. *Solar Energy*, 263, 111920. https://doi.org/10.1016/j.solener.2023.111920
[38] Padmanaban, S., Nasab, M. A., Samavat, T., Nasab, M. A., & Zand, M. (2023). Securing smart power grids against cyber-attacks. In *IoT and Analytics in Renewable Energy Systems* (Vol. 1, pp. 17–36). CRC Press. https://doi.org/10.1201/9781003331117-3
[39] Padmanaban, S., Nasab, M. A., Samavat, T., Zand, M., Nasab, M. A., & Hashemi, E. Cyber security in smart energy networks. In *IoT and Analytics in Renewable Energy Systems* (Vol. 2, pp. 309–325). CRC Press. https://doi.org/1201/9781003374121-26
[40] Lee, C. C., He, Z. W., & Xiao, F. (2022). How does information and communication technology affect renewable energy technology innovation? International evidence. *Renewable Energy*, 200. https://doi.org/10.1016/j.renene.2022.10.015
[41] Tigabu, A. D. (2018). Analyzing the diffusion and adoption of renewable energy technologies in Africa: The functions of innovation systems perspective. *African Journal of Science, Technology, Innovation and Development*, 10(5). https://doi.org/10.1080/20421338.2017.1366130
[42] Goudriaan, Y., Prince, S., & Strzelecka, M. (2023). A narrative approach to the formation of place attachments in landscapes of expanding renewable energy technology. *Landscape Research*, 48(4). https://doi.org/10.1080/01426397.2023.2166911
[43] Shafik, W. (2023). Cyber security perspectives in public spaces: Drone case study. In *Handbook of Research on Cybersecurity Risk in Contemporary Business Systems* (pp. 79–97). IGI Global. https://doi.org/10.4018/978-1-6684-7207-1.ch004
[44] Karatayev, M., Lisiakiewicz, R., Gródek-Szostak, Z., Kotulewicz-Wisińska, K., & Nizamova, M. (2021). The promotion of renewable energy technologies in the former Soviet bloc: Why, how, and with what prospects? In *Energy Reports* (Vol. 7). Elsevier. https://doi.org/10.1016/j.egyr.2021.10.068
[45] Pitelis, A., Vasilakos, N., & Chalvatzis, K. (2020). Fostering innovation in renewable energy technologies: Choice of policy instruments and effectiveness. *Renewable Energy*, 151. https://doi.org/10.1016/j.renene.2019.11.100
[46] Buchmayr, A., Van Ootegem, L., Dewulf, J., & Verhofstadt, E. (2021). Understanding attitudes towards renewable energy technologies and the effect of local experiences. *Energies*, 14(22). https://doi.org/10.3390/en14227596

[47] Ahmed, Z., Adebayo, T. S., Udemba, E. N., Murshed, M., & Kirikkaleli, D. (2022). Effects of economic complexity, economic growth, and renewable energy technology budgets on ecological footprint: The role of democratic accountability. *Environmental Science and Pollution Research*, 29(17). https://doi.org/10.1007/s11356-021-17673-2
[48] Gasser, M., Pezzutto, S., Sparber, W., & Wilczynski, E. (2022). Public research and development funding for renewable energy technologies in Europe: A cross-country analysis. *Sustainability*, 14(9). https://doi.org/10.3390/su14095557
[49] Kachapulula-Mudenda, P., Makashini, L., Malama, A., & Abanda, H. (2018). Review of renewable energy technologies in Zambian households: Capacities and barriers affecting successful deployment. In *Buildings* (Vol. 8, Issue 6). MDPI. https://doi.org/10.3390/buildings8060077
[50] Oryani, B., Koo, Y., Rezania, S., & Shafiee, A. (2021). Barriers to renewable energy technologies penetration: Perspective in Iran. *Renewable Energy*, 174. https://doi.org/10.1016/j.renene.2021.04.052
[51] Edsand, H. E., & Broich, T. (2020). The impact of environmental education on environmental and renewable energy technology awareness: Empirical evidence from Colombia. *International Journal of Science and Mathematics Education*, 18(4). https://doi.org/10.1007/s10763-019-09988-x
[52] Shokoor, F., & Shafik, W. (2023). Harvesting energy overview for sustainable wireless sensor networks. *Journal of Smart Cities and Society*, 2(4). https://doi.org/10.3233/SCS-230016
[53] Suman, A. (2021). Role of renewable energy technologies in climate change adaptation and mitigation: A brief review from Nepal. In *Renewable and Sustainable Energy Reviews* (Vol. 151). Elsevier. https://doi.org/10.1016/j.rser.2021.111524
[54] Moula, M. M. E., Maula, J., Hamdy, M., Fang, T., Jung, N., & Lahdelma, R. (2013). Researching social acceptability of renewable energy technologies in Finland. *International Journal of Sustainable Built Environment*, 2(1). https://doi.org/10.1016/j.ijsbe.2013.10.001
[55] Makki, A. A., & Mosly, I. (2020). Factors affecting public willingness to adopt renewable energy technologies: An exploratory analysis. *Sustainability (Switzerland)*, 12(3). https://doi.org/10.3390/su12030845
[56] Ahmad, M., Ahmed, Z., Gavurova, B., & Oláh, J. (2022). Financial risk, renewable energy technology budgets, and environmental sustainability: Is going green possible? *Frontiers in Environmental Science*, 10. https://doi.org/10.3389/fenvs.2022.909190
[57] Ostapenko, O., Olczak, P., Koval, V., Hren, L., Matuszewska, D., & Postupna, O. (2022). Application of geoinformation systems for assessment of effective integration of renewable energy technologies in the energy sector of Ukraine. *Applied Sciences*, 12(2). https://doi.org/10.3390/app12020592
[58] Sitorus, F., & Brito-Parada, P. R. (2022). The selection of renewable energy technologies using a hybrid subjective and objective multiple criteria decision making method. *Expert Systems with Applications*, 206. https://doi.org/10.1016/j.eswa.2022.117839
[59] Dhirasasna, N. N., & Sahin, O. (2021). A system dynamics model for renewable energy technology adoption of the hotel sector. *Renewable Energy*, 163. https://doi.org/10.1016/j.renene.2020.10.088

[60] Li, D., Heimeriks, G., & Alkemade, F. (2020). The emergence of renewable energy technologies at country level: Relatedness, international knowledge spillovers and domestic energy markets. *Industry and Innovation*, 27(9). https://doi.org/10.1080/13662716.2020.1713734
[61] Shafik, W., Matinkhah, S. M., & Ghasemazade, M. (n.d.). Fog-mobile edge performance evaluation and analysis on internet of things. *Journal of Advance Research in Mobile Computing*, 1(3). https://doi.org/10.5281/zenodo.3591228
[62] Ren, S., Hao, Y., & Wu, H. (2021). Government corruption, market segmentation and renewable energy technology innovation: Evidence from China. *Journal of Environmental Management*, 300. https://doi.org/10.1016/j.jenvman.2021.113686
[63] Ge, T., Cai, X., & Song, X. (2022). How does renewable energy technology innovation affect the upgrading of industrial structure? The moderating effect of green finance. *Renewable Energy*, 197. https://doi.org/10.1016/j.renene.2022.08.046
[64] Li, F., Liu, H., Ma, Y., Xie, X., Wang, Y., & Yang, Y. (2022). Low-carbon spatial differences of renewable energy technologies: Empirical evidence from the Yangtze River Economic Belt. *Technological Forecasting and Social Change*, 183. https://doi.org/10.1016/j.techfore.2022.121897
[65] Shafik, W. (2023). Cyber security perspectives in public spaces: Drone case study. In *Handbook of Research on Cybersecurity Risk in Contemporary Business Systems*. IGI Global. https://doi.org/10.4018/978-1-6684-7207-1.ch004
[66] Shafik, W. (2023). A comprehensive cybersecurity framework for present and future global information technology organizations. In *Effective Cybersecurity Operations for Enterprise-Wide Systems* (pp. 56–79). IGI Global. https://doi.org/10.4018/978-1-6684-9018-1.ch002
[67] Su, Y., & Fan, Q. M. (2022). Renewable energy technology innovation, industrial structure upgrading and green development from the perspective of China's provinces. *Technological Forecasting and Social Change*, 180. https://doi.org/10.1016/j.techfore.2022.121727
[68] Shafik, W., Matinkhah, S. M., & Ghasemzadeh, M. (2020). Internet of things-based energy management, challenges, and solutions in smart cities. *Journal of Communications Technology, Electronics and Computer Science*, 27, 1–11. http://dx.doi.org/10.22385/jctecs.v27i0.302
[69] Wang, W., Xiao, W., & Bai, C. (2022). Can renewable energy technology innovation alleviate energy poverty? Perspective from the marketization level. *Technology in Society*, 68. https://doi.org/10.1016/j.techsoc.2022.101933
[70] Sitorus, F., & Brito-Parada, P. R. (2020). A multiple criteria decision making method to weight the sustainability criteria of renewable energy technologies under uncertainty. *Renewable and Sustainable Energy Reviews*, 127. https://doi.org/10.1016/j.rser.2020.109891
[71] Alnssyan, B., Ahmad, Z., Malela-Majika, J. C., Seong, J. T., & Shafik, W. (2023). On the identifiability and statistical features of a new distributional approach with reliability applications. *AIP Advances*, 13(12). https://doi.org/10.1063/5.0178555

Chapter 2

Soft Computing Techniques in Solar PV Energy Systems

A Review

Sudipta Mohanty, Anita Mohanty, Ambarish G. Mohapatra, Amaresh Gantayat, Subrat Kumar Mohanty, and Sasmita Nayak

2.1 INTRODUCTION

The intersection of soft computing techniques and solar photovoltaic (PV) energy systems presents a captivating area of study, combining advanced computational methods with the imperative goal of harnessing sustainable and clean energy. As societies worldwide increasingly shift towards renewable energy sources, solar PV technology stands out for its inherent eco-friendliness and potential to revolutionize the global energy landscape [1]. This chapter undertakes a thorough review, aiming to illuminate the role, challenges, and advancements of soft computing techniques in the intricate realm of solar PV energy systems.

Solar PV systems, designed to capture energy from the sun, hold immense promise as a key player in the transition to sustainable energy. However, the intermittency and variability inherent in solar energy generation pose significant hurdles to its seamless integration into existing power grids [2]. Soft computing, encompassing a diverse array of computational approaches such as neural networks, fuzzy logic, genetic algorithms, and swarm intelligence, emerges as a strategic solution to address these challenges effectively [3].

This review sets out to explore the diverse applications of soft computing in elevating the performance, efficiency, and reliability of solar PV energy systems. Neural networks, drawing inspiration from the intricate architecture of the human brain, showcase remarkable capabilities in deciphering complex patterns and relationships within solar energy data [4, 5]. Fuzzy logic, with its adaptable decision-making framework, proves invaluable in managing the diverse facets of solar energy systems. Genetic algorithms, mimicking natural evolution, optimize complex systems for the development and management of efficient solar PV setups. Furthermore, swarm intelligence, inspired by collective behaviors observed in social organisms, employs distributed algorithms to optimize various parameters in solar energy systems.

Throughout the review, we delve into the multifaceted dimensions of soft computing techniques, exploring their methodologies and applications in critical areas such as solar resource assessment, energy forecasting, power optimization, fault detection, and seamless grid integration [6–8]. By synthesizing

 DOI: 10.1201/9781003462460-2

existing knowledge and identifying gaps in the current literature, this review aims to provide a comprehensive resource for researchers, practitioners, and policymakers. As we confront the challenges posed by the intermittent nature of solar energy, the integration of soft computing emerges as a promising avenue for achieving a more dependable and effective transition towards sustainable energy solutions, paving the way for a cleaner and greener future.

2.1.1 Solar Photovoltaic Energy System Overview

A solar photovoltaic energy system uses semiconductor materials to generate an electrical current to convert sunlight into power. Due to its favorable effects on the environment and steadily falling prices over time, it is a renewable and sustainable source of energy.

An overview of a solar PV system is shown in Figure 2.1. Its major parts are as follows:

- **Solar Panels (PV Modules):** Solar panels are the most visible part of a PV system. They consist of many individual solar cells made from semiconductor materials, typically silicon. When sunlight hits these cells, it excites electrons, generating direct generating direct current (DC) electricity.

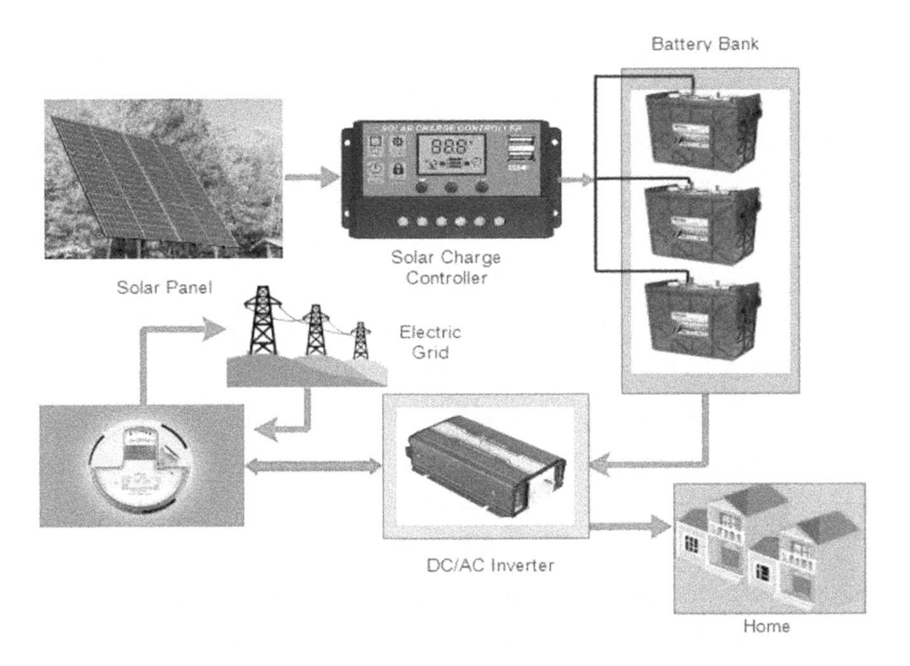

Figure 2.1 An overview of a solar PV system.

- **Inverter:** Alternating current (AC), which is the kind of energy utilized in houses and the power grid, must be created from the DC electricity produced by the solar panels. Therefore, this conversion is made via an inverter.
- **Mounting and Racking:** Solar panels need to be securely mounted on rooftops or the ground. The mounting and tracking system ensures that the panels are properly positioned to capture the maximum amount of sunlight.
- **Solar Charge Controller (for off-grid systems):** In off-grid systems, a charge controller is used to regulate the flow of electricity to the battery bank, preventing overcharging or discharging.
- **Battery bank (for off-grid or hybrid systems):** In some applications, such as remote areas or for backup power, batteries are used to store excess electricity generated during the day for use during the night or when sunlight is not available.
- **Grid Connection (for grid-tied systems):** Most solar PV systems for homes and businesses are grid-tied or linked to the neighbourhood electric grid. It is possible to receive extra power produced throughout the day into the grid in exchange for a credit or payment.
- **Net Metering (for grid-tied systems):** With net metering, companies and homeowners may get paid for the extra power they produce and put back into the grid. They have the option of using the grid as needed during times of poor solar output.
- **Monitoring System:** A monitoring system keeps track of the solar PV system's performance and provides real-time information on energy output and system health. Maintenance and performance optimization are aided by this.
- **Electrical Panel (Load Centre):** The electrical panel of the building receives the power produced by the solar PV system. This power is used by the structure to operate electrical equipment like lights, appliances, and other things.
- **Backup Generator (for off-grid systems):** A backup generator can be used in conjunction with solar PV and battery storage to guarantee a constant power supply in isolated places or for important applications.

Solar PV systems need little maintenance. Typical panel care includes

- Regular cleaning
- Monitoring for dirt or shading
- Occasional inverter upkeep.

Solar PV systems have the following advantages, such as

- **Environmental Benefits:** Solar PV systems produce electricity without emitting greenhouse gases or air pollutants, making them an environmentally friendly energy source. They contribute to reducing reliance on fossil fuels and mitigating climate change.

- **Financial Incentives:** Solar photovoltaic systems generate power without releasing greenhouse gases or other air pollutants, making them a green energy source. They help mitigate climate change by lowering dependency on fossil fuels.
- **Longevity:** Solar panels normally last for 25 to 30 years or more, and as time passes, their efficiency steadily declines.

Solar PV technology continues to evolve, with advancements in materials and design leading to increased efficiency and reduced costs. This makes solar energy an increasingly attractive option for both residential and commercial applications as we move toward a more sustainable energy future.

2.1.2 Importance of Soft Computing in Solar PV Systems

The efficiency, dependability, and overall performance of solar photovoltaic systems are significantly improved by soft computing [9]. Here are a few significant factors as shown in Figure 2.2 which emphasize the use of soft computing in solar PV systems:

- **Maximum Power Point Tracking (MPPT):** Advanced MPPT algorithms are implemented using soft computing methods like fuzzy logic and neural networks. These algorithms enable the solar PV system to continuously modify its operating point to capture the most energy from various environmental factors, such as variations in temperature and sunlight intensity.
- **Prediction and Forecasting:** To create precise short-term and long-term forecasts, soft computing techniques, notably artificial neural networks (ANNs) and machine learning algorithms, may analyze historical meteorological data, solar irradiance trends, and system performance. These projections aid grid administrators and operators in more efficiently planning grid integration, scheduling maintenance, and anticipating energy generation trends.
- **Fault Detection and Diagnostics:** Real-time fault identification and diagnosis in solar PV systems use soft computing approaches. These techniques can recognize and localize problems like module deterioration, inverter failures, or shading concerns by analyzing sensor data and system characteristics. Early detection helps cut down on maintenance expenses and downtime.
- **Energy Management:** Soft computing techniques may be used to optimize the performance of solar PV installations both on and off the grid. These algorithms support the control of energy flows, the prioritization of load utilization, and the choice of whether to store extra energy in batteries or return it to the grid. The dependability and economic feasibility of solar energy systems are enhanced by this.

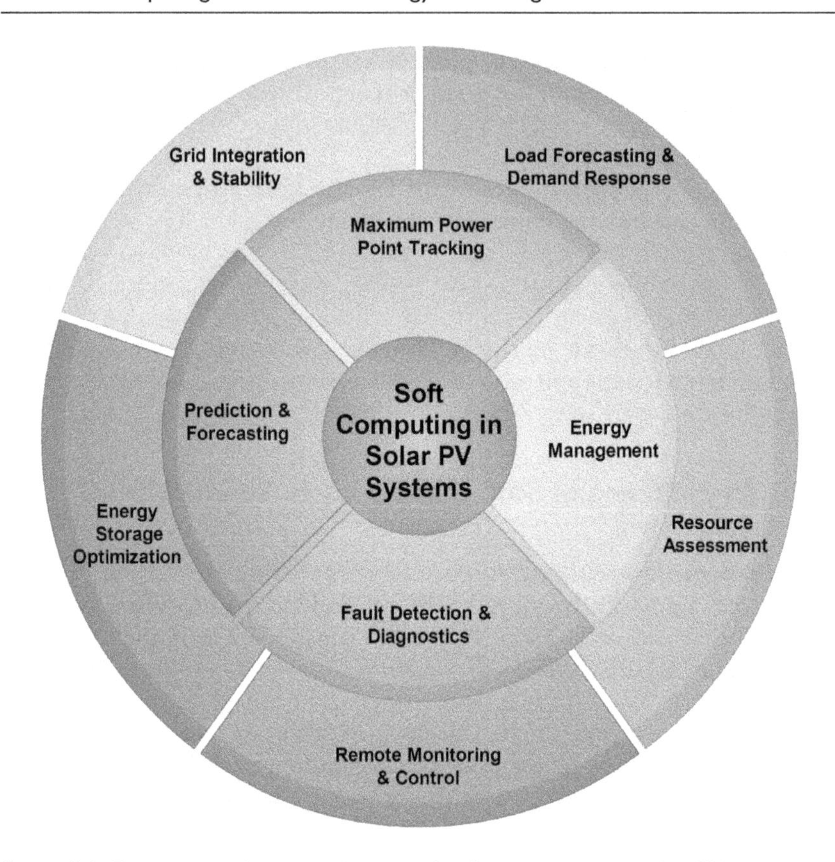

Figure 2.2 Factors emphasizing the use of soft computing in solar PV systems.

- **Load Forecasting and Demand Response:** Soft computing can help with load forecasting in grid-connected PV systems, which supports grid operators in balancing supply and demand. Additionally, it can help demand response methods by foreseeing the times when there will be extra solar energy that can be put to use for things like heating water or charging electric vehicles.
- **Grid Integration and Stability:** Soft computing methods can help maintain grid stability as the grid's penetration of solar PV rises. To maintain a steady and stable energy supply, advanced control algorithms can assist in regulating the variability and intermittency of solar power.
- **Energy Storage Optimization:** The performance of energy storage devices (like batteries) in solar PV installations may be optimized via soft computing. These approaches decide when to charge and discharge batteries for optimal effectiveness and cost savings by examining previous data and current conditions.

- **Resource Assessment:** By examining past weather and solar irradiance data, soft computing can help with site selection for future solar PV systems. This aids in locating areas with the greatest potential for energy generation.
- **Remote Monitoring and Control:** Soft computing makes it possible to remotely monitor and manage solar PV systems, giving operators the ability to adapt to shifting circumstances and make modifications in real-time without having to physically intervene.

The soft computing approaches improve solar PV system efficiency, performance, and cost-effectiveness by enabling intelligent control, forecasting, and optimization. The significance of soft computing in controlling and optimizing these systems grows as our energy infrastructure develops to incorporate renewable energy sources like solar PV.

2.2 SOFT COMPUTING TECHNIQUES

A collection of computer methods and approaches known as "soft computing" are modeled after how the human mind can reason and choose in the face of ambiguity and imprecision. Soft computing approaches are used to solve complicated, real-world issues where ambiguity, uncertainty, and incomplete information are frequent. These problems are handled differently from traditional "hard" computing, which is based on exact mathematical models and binary logic.

The most important soft computing methods are shown in Figure 2.3 and are listed as

- Fuzzy logic
- Artificial neural networks (ANNs)
- Genetic algorithms (GAs)
- Particle swarm optimization (PSO)
- Ant colony optimization (ACO)
- Support vector machines (SVMs)
- Bayesian networks
- Rough sets
- Swarm intelligence
- Neuro-fuzzy systems.

Soft computing approaches are especially useful in fields like artificial intelligence, control systems, data mining, robotics, and decision support systems when real-world data is noisy, imperfect, or ambiguous. They give problem-solving tools that are adaptable and versatile, enabling computers to handle challenging, human-like reasoning tasks.

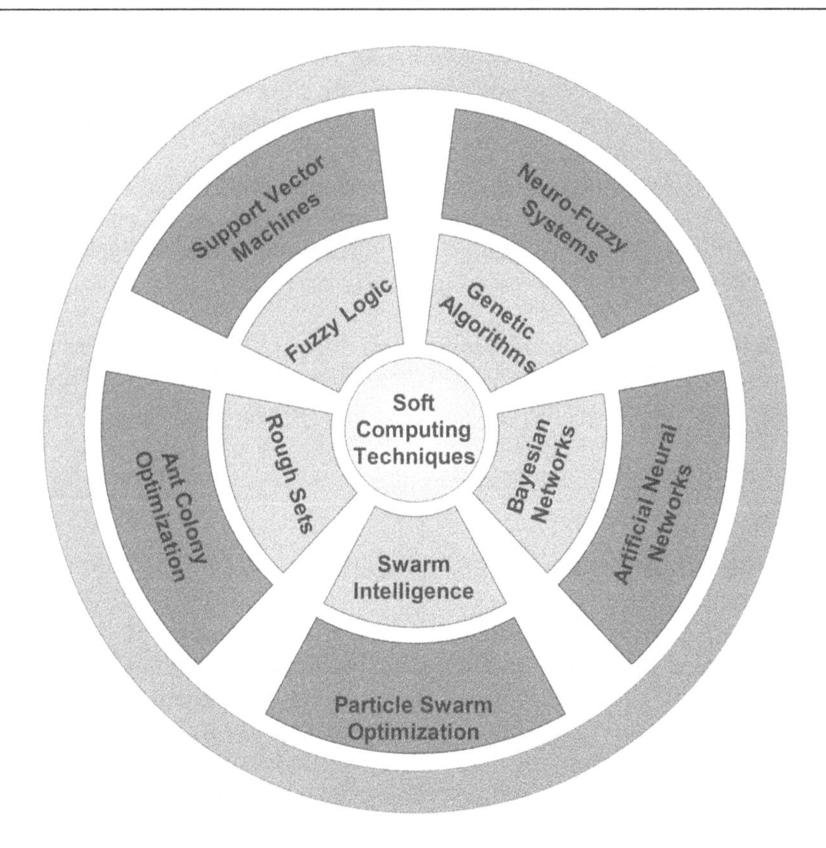

Figure 2.3 Different soft computing techniques.

2.2.1 Neural Networks for Solar Energy Prediction

Neural networks can capture the complicated, nonlinear interactions between the many variables that impact solar irradiance and, in turn, energy output; hence they are an effective tool for forecasting solar energy generation. Here are some examples of how neural networks may be used to forecast solar energy production:

- **Data collection:** Historical data on solar irradiance, meteorological conditions (such as temperature and cloud cover), and solar energy production at the site of interest are required to create an accurate prediction model. Weather stations, satellites, or local sensors may provide this data.
- **Data Processing:** There are frequent noise and outliers in raw data frequently. To guarantee that neural networks can function properly, data pre-treatment operations include cleaning the data, addressing missing values, and normalizing or scaling the data.

- **Feature Selection:** Select pertinent characteristics (variables) that affect the production of solar energy. Latitude, longitude, season, and meteorological information (such as temperature, humidity, and cloud cover) may all play a role in this. To increase prediction accuracy, feature engineering may entail developing new features or modifying current ones.
- **Neural Network Architecture:** Create the architecture for the neural network. A feed-forward neural network (FNN) or recurrent neural network (RNN), like long short-term memory (LSTM), is frequently used for solar energy prediction. The decision is based on the type of data and the nature of the prediction task.
- **Training Data:** Create training and validation datasets from the historical data. The neural network is taught to understand the correlations between input characteristics, such as weather data, and the goal variable, solar energy generation, using the training dataset. The validation dataset assists in preventing overfitting and monitoring the network's effectiveness during training.
- **Model Training:** To train the neural network, use the training dataset. The network modifies its weights and biases during training to reduce the discrepancy between its forecasts and the actual solar energy output. This procedure is repeated until the model reaches an acceptable degree of accuracy.
- **Model Evaluation:** The neural network should be trained using the training dataset. To reduce the discrepancy between its forecasts and the actual solar energy output, the network modifies its weights and biases throughout training. This procedure is repeated until the model's accuracy converges to an acceptable level.
- **Deployment:** The neural network may be used for real-time or almost real-time solar energy prediction once it has been trained and verified. It offers an estimate of solar energy generation as an output and accepts current or predicted meteorological data as input.
- **Continuous Monitoring and Updating:** Models for predicting solar energy should be updated often with fresh information to account for shifting climatic circumstances. Prediction accuracy is maintained by ongoing monitoring.
- **Post-Processing:** Predictions can be improved using post-processing methods. For instance, you may employ filtering methods or smoothing algorithms to lessen sudden changes in the expected energy production.
- **Integration with Control System:** To optimize energy usage and grid integration, predicted solar energy generation may be included in energy management and control systems.

Given their adaptability and capacity to simulate intricate correlations in solar energy data, neural networks are an important tool for increasing the efficiency and dependability of solar energy systems as well as for grid management and energy planning.

2.2.2 Fuzzy Logic Control for Maximum Power Point Tracking

Maximum power point tracking, or MPPT for short, describes how to get the most power out of a solar panel under any set of environmental conditions, including shade, temperature, and radiation. This objective can be accomplished in several ways, but the general procedure involves proving the maximum power transfer theorem, one of the foundational principles of electrical circuits. This principle states that when the resistances of the source and the load are equal, maximum power transmission from the source to the load occurs. Maximal power feed to the load can be efficiently ensured by applying the maximum power theorem. This is accomplished in solar systems using a controller and a converter.

A fuzzy logic controller (FLC) [10] can be used in a wide range of applications involving renewable energy resources. Because FLC is so simple, its requirements have grown over the past 10 years. Additionally, FLC handles imprecise input, for which the controller does not require an accurate mathematical model. Nonlinearity circumstances are easily handled by FLC to maximize the power output from PV modules [11, 12]. It can function in any type of weather and adapt to variations in temperature and light intensity.

The fuzzy logic controller is proposed in this chapter as a MPPT system to get maximum power from photovoltaic with changes in irradiation and temperature. By continually altering the operating point to utilize the maximum amount of electricity from the solar panels, MPPT is crucial for ensuring that a solar PV system performs at its highest efficiency. The flow chart shown in Figure 2.4 shows the steps of utilization of fuzzy logic control in MPPT.

Fuzzy logic control offers several advantages for MPPT in solar PV systems:

- **Robustness:** Fuzzy logic can handle the nonlinear and dynamic nature of PV systems, making it robust in various environmental conditions.
- **Adaptability:** Fuzzy logic controllers can adapt to changing conditions without needing a precise model of the system.
- **Simplicity:** FLCs are relatively easy to implement and tune compared to some other MPPT algorithms.
- **Improved Efficiency:** By continuously tracking the maximum power point, FLCs help maximize energy harvesting from the solar panels.

However, the performance of a FLC depends on the proper selection of membership functions, rule base, and tuning of parameters to match the specific characteristics of the PV system and environmental conditions.

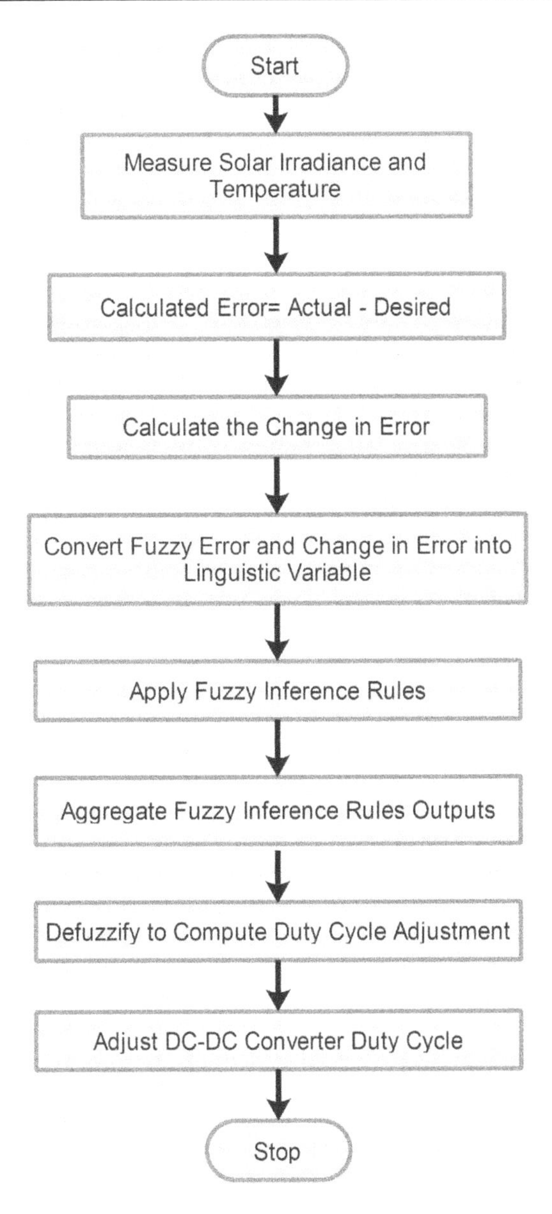

Figure 2.4 Flow chart for steps of utilization of fuzzy logic control in MPPT.

2.2.3 Genetic Algorithms for Parameter Optimization

Genetic algorithms (GAs) are powerful optimization techniques inspired by the process of natural selection and evolution. They can be applied to various optimization problems, including parameter optimization. A broad description of how genetic algorithms can be applied to parameter optimization is given next:

- Create an initial population of potential solutions (parameter sets).
- Evaluate the fitness of each solution in the population.
- Select solutions from the current population to serve as parents for the next generation.
- Create new solutions (offspring) by combining the genetic material of two parent solutions.
- Introduce small random changes into some of the parameter values of the offspring.
- Replace some solutions in the current population with the newly created offspring.
- Decide when to stop the optimization process.
- The best solution found throughout the optimization process represents the optimized set of parameters.

Genetic algorithms are effective for parameter optimization for several reasons:

- **Global Search:** GAs are well-suited for problems with complex, multidimensional search spaces and multiple local optima. They explore a wide range of solutions and have the potential to find global optima.
- **Adaptability:** GAs can adapt to changing conditions and provide robust solutions. Mutations introduce random changes, allowing the algorithm to escape local optima.
- **Parallelism:** GAs can be parallelized, which means that multiple solutions can be evaluated simultaneously, making them suitable for high-performance computing environments.
- **No Derivative Information:** GAs do not require derivatives or gradient information, which is advantageous for problems where derivatives are unavailable or difficult to compute.

2.2.4 Swarm Intelligence for Energy Management

A subfield of artificial intelligence known as swarm intelligence (SI) draws its inspiration from the group behavior of social insects and other animal species. SI algorithms simulate how members of a group work together to accomplish difficult tasks. When used in energy management, SI may optimize the distribution and use of energy resources, boost system dependability, and increase efficiency. The following are some applications of swarm intelligence in energy management:

- Demand response optimization
- Microgrid control
- Smart grid operation
- Energy storage management
- Renewable energy forecasting
- Load balancing and optimization
- Energy efficient routing
- Energy efficient building control
- Energy trading in peer-to-peer networks
- Fault detection and diagnostics
- Energy-aware manufacturing.

2.3 SOLAR RESOURCE ASSESSMENT

A critical step in assessing the viability and potential of solar energy projects is solar resource evaluation. It entails the measurement and analysis of several elements connected to geographic location, weather, and solar radiation. For project developers, investors, and policymakers to make educated choices about using solar energy in a particular area, solar resource assessment is an important phase in the development of solar energy projects. For optimizing the design and performance of solar PV systems and maintaining the financial feasibility of solar projects, accurate assessments are crucial [13].

2.3.1 Soft Computing-Based Solar Irradiance Prediction Models

Due to their capacity to manage the complexity and uncertainty involved with meteorological and environmental data, soft computing-based models are increasingly being utilized to estimate solar irradiance [14]. These models increase the precision of solar irradiance predictions by utilizing methods such as fuzzy logic, neural networks, and genetic algorithms. Here are some typical soft computing techniques for predicting solar irradiance:

- **Fuzzy Logic-Based Models:**
  Fuzzy Inference Systems (FIS): The language factors and laws regulating sun irradiance patterns can be captured using fuzzy logic. To forecast solar irradiance levels, FIS models include elements including cloud cover, the time of day, and previous data. Uncertainty and linguistic words are represented by fuzzy membership functions.

- **Neural Network-Based Models:**
  Artificial Neural Networks (ANNs): To generate predictions, ANNs may understand intricate associations in past data. To anticipate future solar irradiance levels, artificial neural networks use inputs such as meteorological conditions (temperature, humidity, cloud cover).
  Recurrent Neural Networks (RNNs): For time series data like sun irradiance, RNNs, particularly long short-term memory (LSTM) networks, are ideally suited. They can identify patterns and temporal relationships in historical irradiance data.
- **Hybrid Models:** Prediction accuracy may be improved by combining several soft computing approaches. For instance, a hybrid model may pre-process input using fuzzy logic before feeding it to a neural network for the final prediction.
- **Genetic Algorithms:** The parameters of solar irradiance prediction models may be optimized using GAs. They aid in fine-tuning the model's hyperparameters and architecture to increase forecasting precision. GAs can be used with other soft computing methodologies.
- **Ensemble Models:** To increase accuracy and decrease mistakes, ensemble approaches aggregate predictions from many models. This method can be very useful for soft computing-based predictions of solar irradiation.
- **Data Fusion:** To improve the precision of estimates of solar irradiance, soft computing models can incorporate data from many sources, including satellite photos, ground-based sensors, and weather forecasts. Techniques for data fusion assist in combining data from many sources.
- **Explanatory Models:** Some soft computing models offer comprehensible results that let consumers comprehend the elements influencing estimates of solar irradiation. Both system analysis and decision-making may benefit from this.
- **Online Learning:** Soft computing models may be generated for online learning, enabling them to change and update forecasts in real-time as fresh data becomes available. This is vital in the case of changing weather conditions.
- **Uncertainty Quantification:** Soft computing models can give probability distributions or confidence ranges for future irradiance levels, which may be used to enumerate forecast uncertainty. For risk assessment and decision-making, this knowledge is supportive.
- **Spatial Prediction:** Soft computing models may be expanded to forecast solar irradiance at various places, which makes it easier to build and manage distributed solar energy systems.

2.3.2 Cloud Cover and Solar Energy Prediction Using Neural Networks

The intricate and dynamic structure of cloud patterns makes it difficult to predict solar energy output when there is fluctuating cloud cover. Modeling the link between cloud cover and solar energy output may be done well using neural networks, especially recurrent neural networks (RNNs) and convolution neural networks (CNNs) [15]. Here's how you can use neural networks for cloud cover and solar energy prediction:

- Gather historical data on solar energy generation and corresponding weather conditions, including cloud cover.
- Clean and pre-process the data by handling missing values, normalizing data, and converting it into a suitable format for neural network training.
- Extract relevant features from the data that can help the neural network model capture the relationship between cloud cover and solar energy production.
- Split the dataset into training, validation, and testing sets.
- Choose the appropriate neural network architecture for the task.
- Train the neural network on the training dataset.
- Optimize the model's hyperparameters, including the number of layers, the number of neurons in each layer, learning rate, batch size, and dropout rates.
- Evaluate the trained neural network on the testing dataset using appropriate evaluation metrics, such as mean squared error (MSE), root mean squared error (RMSE), or coefficient of determination (R-squared).
- Depending on the specific application, you may need to post-process the model's predictions.
- Deploy the trained neural network model for real-time solar energy prediction.
- Continuously monitor the model's performance and update it as new data becomes available.

With the ability to understand intricate, nonlinear correlations between cloud cover and solar energy output, neural networks are advantageous as instruments for precise solar energy forecasting under real-world conditions.

2.3.3 Fuzzy Logic Approaches for Solar Resource Assessment

By offering a flexible and adaptive framework for addressing uncertainty, imprecision, and complicated linkages in the data, fuzzy logic techniques might help evaluate solar resource potential. Table 2.1 shows the different ways how fuzzy logic may be used in the context of evaluating solar resources.

Overall, fuzzy logic methods provide a flexible and understandable framework for evaluating solar resources, enabling the integration of various data sources and specialist expertise. They can aid in enhancing the precision and dependability of solar energy forecasts, which is crucial for efficient energy management and planning.

Table 2.1 Different Ways of Evaluation of Solar Resources by Fuzzy Logic

Different Ways	*Solar Resource Evaluation*
Data Fusion	Fuzzy logic can combine data from several sources, including weather predictions, satellite observations, and data from ground-based sensors, to produce a more thorough and precise evaluation of solar irradiance. When working with data from several and perhaps conflicting sources, this is especially helpful.
Quality Control	Fuzzy logic can be employed for quality control of solar radiation data. It can assist in locating and removing inaccurate or outlier measurements brought on by broken sensors or severe weather.
Temporal Data Smoothing	Smoothing temporal fluctuations in solar irradiance data with fuzzy logic can help to lower noise and make it simpler to spot long-term trends and patterns.
Irradiance Forecasting	By using linguistic variables and rules that reflect the effect of elements like cloud cover; time of day, and historical data, fuzzy logic can improve short-term and long-term solar irradiance forecasting models.
Resource Classification	Based on their potential for solar energy, locations or areas can be grouped using fuzzy logic. When choosing a location for solar energy projects, this categorization might be useful.
Cloud Cover Assessment	Cloud cover may be measured and evaluated using fuzzy variables using fuzzy logic, allowing for a more detailed depiction of cloudiness levels and how they affect solar irradiance.
Uncertainty Quantification	The degree of uncertainty in the findings of solar resource assessments may be estimated using fuzzy logic. This is useful for explaining to decision-makers and stakeholders how reliable predictions are.
Incorporating Expert Knowledge	Experts can include their subject knowledge and expertise in the evaluation process by using fuzzy logic. The definition of membership functions and fuzzy rules that direct the assessment may be done using this information.
Visualization	Using linguistic concepts and fuzzy sets, fuzzy logic may create visual representations of data about solar resource availability. Non-experts can better grasp the assessment results with the aid of these visualizations.
Adaptability	Fuzzy logic models are appropriate for dynamic solar resource assessment scenarios because they can adjust to changing environmental circumstances and data availability.
Hybrid Models	Other modeling methods, such as neural networks or evolutionary algorithms, can be combined with fuzzy logic to build hybrid models that make use of the best aspects of each method.
Decision Support	By evaluating the feasibility of a place for solar energy projects based on a variety of factors, such as solar resource potential, land use, and regulatory constraints, fuzzy logic can help with decision-making.

2.4 ENERGY PREDICTION AND OPTIMIZATION

Modern energy management systems must include energy forecasting and optimization, especially when using renewable energy sources and sustainable resource management [16]. These procedures entail predicting energy production or consumption, followed by the optimization of energy-related operations to meet predetermined objectives.

Real-time control systems, machine learning, linear programming, mixed-integer programming, and other optimization methods are frequently used in energy forecasting and optimization strategies. They are essential for attaining energy efficiency, lowering carbon emissions, and providing a steady and sustainable supply of energy.

2.4.1 Load and Solar Energy Prediction with Neural Networks

For effective grid management, demand response, and renewable energy integration, neural networks must be used to predict both energy demand (load) and solar energy output. To describe the intricate linkages and temporal correlations in these data streams, neural networks, such as recurrent neural networks (RNNs) and feed-forward neural networks (FNNs), can be employed. Here is a way to forecast solar energy and load using neural networks:

Data Collection and Pre-Processing:

- Data collection: Assemble historical information on solar energy production and energy load (demand). Time-stamped measurements of the load, solar irradiance, and other pertinent variables (such as temperature and cloud cover) should all be included in this data.
- Data Processing: By addressing missing values, smoothing noisy data, and normalizing the data to a common scale, you may clean and pre-process the data. Make sure the data is timely and well-aligned.

Load Prediction:

- Feature Engineering: For load prediction, choose pertinent features. These might consist of past load data, time and date details, weather information, and any other elements that are known to affect energy consumption.
- Neural Network Architecture: Create a neural network design that is appropriate for predicting load. It is typical to utilize a recurrent neural network, such as a LSTM network, or a feed-forward neural network. Tasks requiring sequence prediction, such as time series forecasting, are well-suited for LSTM networks.

- Training Data: Create training, validation, and testing datasets from the historical data. The validation dataset aids in hyperparameter tuning, the testing dataset is utilized for assessment, and the training dataset is used to train the neural network.
- Model Training: Utilize the training dataset to train the neural network. The network gains the ability to recognize relationships and trends in past load data.
- Hyperparameter Tuning: To improve the model's performance on the validation data set, try out various network designs, learning rates, batch sizes, and regularization methods.
- Load Prediction: Utilize the trained neural network to anticipate the load for the next time steps. A series of historical load data is used as the network's input, and the expected future load is produced as its output.

Solar Energy Prediction:

- Feature Engineering: For the forecast of solar energy, choose pertinent features. These generally include information on past solar irradiance, the time of day, the season, and maybe weather conditions like cloud cover.
- Neural Network Architecture: Create a neural network design that is appropriate for predicting solar energy. LSTM networks are frequently used for time series forecasting of solar energy generation, much like load prediction.
- Training Data: Create training, validation, and testing datasets using historical data on solar energy production.
- Model Training: Utilize the training dataset to train the neural network. The network gains the ability to depict the connection between solar irradiance and energy production.
- Solar Energy Prediction: Using the input characteristics, such as historical irradiance data and other pertinent elements, the trained neural network may be used to forecast future solar energy generation.

Evaluation and Deployment:

- Model Evaluation: Use relevant metrics, such as mean squared error (MSE), root mean squared error (RMSE), or coefficient of determination (R-squared), to assess the performance of both the load and solar energy prediction models on the test datasets.
- Deployment: Use the trained neural network models to estimate load and solar energy in real-time or almost real-time applications. These forecasts may be included in platforms for energy trade, grid operations, and energy management.

2.4.2 Fuzzy Logic-Based Energy Management Systems

Fuzzy logic-based energy management systems (FLEMS) offer a comprehensive method for streamlining energy distribution and consumption in a variety of settings, from smart buildings to industrial processes to the incorporation of renewable energy sources. To deal with the uncertainty and imprecision that are inherent in energy systems; FLEMS uses the concepts of fuzzy logic. To improve energy efficiency and sustainability, FLEMS can model complicated interactions and make judgments in real-time by using linguistic variables, fuzzy membership functions, and a rule-based inference system.

The flexibility of FLEMS to adapt to changing circumstances is a noteworthy quality. They continually assess incoming information, such as energy demand, generation, and environmental conditions, and modify system settings as necessary. This versatility enables FLEMS to quickly react to altering energy demands, variations in renewable energy output, and unanticipated disturbances, assuring reliable system operation and efficient energy consumption.

FLEMS can optimize the performance of equipment and processes in industrial environments, reducing energy waste while maintaining production goals. In intelligent buildings, they control HVAC, lighting, and appliances according to user preferences, weather, and occupancy, resulting in energy savings and improved occupant comfort. When it comes to integrating renewable energy, FLEMS oversees the distribution of electricity from various sources, such as solar and wind, in a way that maximizes the use of renewable energy while preserving grid stability.

2.4.3 Genetic Algorithms for PV System Sizing and Configuration

Photovoltaic (PV) system sizing and design may be optimized using genetic algorithms (GAs), making the systems more effective and economical. To determine the ideal combination of PV system characteristics, GAs imitate the concepts of natural selection and evolution in the context of PV system design [17].

The first step in the procedure is to define a fitness function, which stands for the optimization's goal, which might range from maximizing energy production to minimizing system costs. In the population of the genetic algorithm, parameters like the number and kind of PV panels, the size of the inverter, the battery capacity, and tilt angles are regarded as genes. A population of potential answers is initially produced at random or utilizing pre-existing data.

GAs assess the fitness of each solution using repeated generations and the fitness function. It is more likely that solutions that perform better about the optimization target will be chosen for replication. A new generation of solutions is created during the reproduction phase by combining (crossover) or randomly altering (mutation) existing solutions.

GAs continue to hone and enhance the solutions across succeeding generations. For the PV system, this evolutionary process frequently converges on optimal or nearly optimal designs. When a predetermined termination requirement is satisfied, such as after a predetermined number of generations or when the fitness function exceeds a particular threshold, the algorithm terminates.

The benefits of employing GAs for PV system setup and sizing lay in their capacity to effectively explore a large solution space while considering complicated interactions between system components and restrictions. By doing this, GAs can uncover solutions that might be difficult to locate using conventional manual design or laborious search techniques. This feature is especially useful as distributed generation networks and energy grids increasingly integrate PV systems, which need optimal setup to maximize their influence on energy production and consumption. As a result, GAs provide a useful method for creating economical and efficient PV systems, promoting a wider use of solar energy technology.

2.5 FAULT DETECTION AND DIAGNOSTICS

Maintaining the performance and dependability of complex systems, such as industrial machinery, mechanical systems, electrical circuits, and software applications, requires the use of fault detection and diagnostics (FDD). FDD is a generic term for a variety of procedures and techniques used to uncover anomalous behavior, flaws, or malfunctions in a system and determine their underlying causes.

FDD's main objective is to find differences from anticipated or typical system behavior, which might be a sign of flaws or abnormalities. These variations may appear as anomalous data patterns, sensor readings, performance measures, or error messages. To analyze this data, FDD systems use a variety of data-driven and model-based techniques.

2.5.1 Anomaly Detection in PV Arrays Using Neural Networks

The dependable operation and performance of solar energy systems are now ensured by anomaly detection in photovoltaic (PV) arrays utilizing neural networks. Deep learning neural network models provide a reliable method for spotting anomalous behavior and possible problems in PV arrays [18].

The first step in the procedure is data collecting, which involves continually monitoring a variety of sensor measures from the PV array, including solar irradiance, temperature, voltage, and current. These data streams offer insightful information on the functionality and health of the system.

Neural networks are then trained using historical data, where both normal and faulty operating conditions are represented. To discern between normal behavior and deviations brought on by abnormalities or errors, the network learns the patterns and correlations present in the data. The design of the neural network may be altered to account for the complexity of the data and the unique features of the PV system.

The operational trained neural network continually and instantly analyses incoming sensor data. The neural network sends out an alert or message when it notices a change from the expected behavior, enabling maintenance teams to act quickly. Shade, soiling, partial module failures, or wiring problems are often seen as abnormalities that might affect the PV array's energy production.

Neural networks are advantageous for PV array anomaly detection because they can handle complicated, high-dimensional data and capture nonlinear correlations. They are excellent at spotting subtle patterns that rule-based or conventional statistical approaches would find difficult to detect. Additionally, neural networks are well-suited for dynamic, real-world contexts because of their capacity to adapt to changing circumstances and develop fault scenarios.

2.5.2 Fuzzy Logic-Based Fault Detection in PV Inverters

In the context of photovoltaic (PV) inverters, fuzzy logic-based fault detection is a useful method that improves the performance and dependability of solar energy systems. Fuzzy logic, which is excellent at handling uncertainty and imprecision, is used in this method to find abnormalities and problems in PV inverters.

Data from numerous sensors and monitoring devices installed within the PV inverter system are first gathered and processed. Critical characteristics including voltage, current, temperature, and frequency are measured by these sensors. The behavior of the operational parameters is then represented by linguistic variables and membership functions using fuzzy logic-based defect detection. The system can function with erroneous data and subjective descriptions because of its language representation.

The links between the language variables and potential fault situations are defined by a rule-based framework that integrates expert knowledge and heuristics. The decision-making process for defect detection is encapsulated in these rules, which are expressed as "if-then" statements. When fresh data is gathered, it goes through a process called fuzzification where defined membership functions are used to transform exact numerical values into fuzzy sets. The system can now incorporate data variances and uncertainties thanks to this modification.

The specified rules are then applied to the fuzzified data by the fuzzy logic inference engine, which concludes the PV inverter's present operating condition. Based on the fuzzy logic membership functions, it evaluates the extent to which each rule is followed. Using aggregation operators, the individual rule findings are combined to create a comprehensive evaluation of the state of the system.

To offer a clear indicator of whether a fault or abnormality is found and, if so, the severity of the problem, the fuzzy output is finally de-fuzzified. Depending on the results, the system may sound alerts, start remedial operations, or give maintenance staff diagnostic data. This knowledge, which may include the fault's kind and location, enables swift and precise treatments.

2.6 GRID INTEGRATION AND SMART SYSTEMS

Grid integration and smart systems play crucial roles in optimizing the performance and efficiency of solar PV (photovoltaic) energy systems [19–21]. The integration of solar PV into the electrical grid requires advanced technologies and intelligent systems to ensure a reliable and stable energy supply. Here it explores the key aspects of grid integration and smart systems in the context of this topic:

- **Grid Integration:**
  - Challenges: Solar PV systems are intermittent and dependent on weather conditions, which pose challenges for grid integration. Fluctuations in solar power generation can lead to issues like voltage instability, frequency variations, and grid imbalances.
  - Smart Inverters: Grid integration is facilitated using smart inverters. These inverters incorporate advanced control algorithms to regulate the flow of electricity, ensuring a smooth and stable connection to the grid. They can also provide grid support functions, such as reactive power control and voltage regulation.
  - Energy Storage Systems: To address the intermittency of solar power, energy storage systems, such as batteries, can be integrated. Smart control strategies optimize the charging and discharging of these storage systems to balance the supply and demand of electricity on the grid.

- **Smart Systems in Solar PV Energy:**
  - Monitoring and Control: Soft computing techniques, such as fuzzy logic and neural networks, can be employed for real-time monitoring and control of solar PV systems. These methods enable adaptive and intelligent decision-making based on the current operating conditions.
  - Predictive Analytics: Soft computing models can be used for predicting solar power generation based on weather forecasts,

historical data, and other relevant parameters. This helps in better planning and grid management, allowing utilities to anticipate and mitigate potential issues.
- Fault Detection and Diagnostics: Soft computing techniques contribute to fault detection and diagnostics in solar PV systems. By analyzing performance data using intelligent algorithms, potential issues can be identified early, leading to improved system reliability and reduced downtime.

- **Optimization of Energy Output:**
  - Maximum Power Point Tracking (MPPT): Soft computing algorithms, such as genetic algorithms and particle swarm optimization, can enhance MPPT techniques. These methods optimize the operation of solar PV panels to extract the maximum available power under varying environmental conditions.
  - Load Forecasting: Soft computing models can aid in load forecasting, allowing for better management of solar power generation and consumption. This helps utilities plan for peak demand periods and optimize grid operations.

- **Integration of Soft Computing Techniques:**
  - Hybrid Approaches: Combining different soft computing techniques, such as fuzzy logic and genetic algorithms, can result in hybrid models that offer improved accuracy and robustness in addressing the complex dynamics of solar PV energy systems.

In summary, the integration of soft computing techniques in solar PV energy systems enhances grid integration and smart system functionalities. These intelligent approaches contribute to improved efficiency, reliability, and overall performance of solar power generation, making it a valuable area of research and application in the renewable energy domain.

2.6.1 Soft Computing Approaches for Grid Integration

When it comes to accepting renewable energy sources, improving energy efficiency, and assuring grid stability, soft computing methods are becoming more and more crucial to the challenging work of grid integration. Grid operators are looking to fuzzy logic, neural networks, genetic algorithms, and swarm intelligence for answers to the unpredictability and fluctuation of renewable energy sources like solar and wind. These soft computing methods are excellent at managing complexity and ambiguity.

For example, neural networks are used to anticipate renewable energy production accurately utilizing both historical data and real-time inputs, such as weather. Grid managers can effectively manage the energy supply thanks to this. For efficient power flow, genetic algorithms are used to help maintain a smooth balance between supply and demand, which is essential in a grid that is becoming more decentralized.

Systems based on fuzzy logic are useful for fault detection and diagnostics because they can quickly spot abnormalities or disturbances in real-time. This swift reaction ensures grid reliability and reduces interruptions. Additionally, soft computing strategies play a crucial role in managing distributed energy resources, controlling microgrids, and improving energy storage systems, all of which improve the resilience and flexibility of the grid.

2.6.2 Swarm Intelligence-Based Microgrid Management

Modern methods for maximizing the use and administration of distributed energy resources within a microgrid ecosystem include swarm intelligence-based microgrid management [22]. Swarm intelligence algorithms, which are modelled after natural systems like ant colonies and flocks of birds, enable autonomous decision-making among various grid components, including solar panels, wind turbines, energy storage devices, and loads. To achieve shared goals like load balancing, energy optimization, and grid stability, each component in this paradigm is viewed as an autonomous agent that interacts and cooperates with others.

To coordinate these creatures, swarm intelligence techniques like particle swarm optimization (PSO) and ant colony optimization (ACO) are used. PSO, for instance, represents agents as particles that repeatedly modify their behavior depending on both their individual experiences and the swarm's knowledge. ACO, on the other hand, uses virtual pheromones that are deposited by grid components to express their status and exchange information, much as ants use pheromone trails to discover the best pathways.

This strategy has various benefits for managing microgrids. It is appropriate for dynamic situations with variable renewable energy supply and varying load needs first because it encourages scaling and adaptation. Empowering agents to self-organize and adjust to unforeseen occurrences or failures also improves grid resilience and guarantees a constant supply of electricity. Swarm intelligence-based management also optimizes resource distribution, enhancing cost- and energy-effectiveness.

Overall, decentralized, autonomous, and adaptive grid operating concepts are in line with swarm intelligence-based microgrid management. Utilizing distributed agents' collective intelligence improves microgrid performance while also aiding in the transition to larger energy systems that are more robust, sustainable, and effective.

2.7 CASE STUDIES AND APPLICATIONS

2.7.1 Case Study 1: Neural Network-Based Solar Energy Forecasting

2.7.1.1 Introduction

An essential part of grid integration techniques, neural network-based solar energy forecasting enables grid operators to efficiently manage the fluctuation of renewable energy sources like solar electricity. This case study investigates the implementation of a neural network-based solar energy forecasting system by a utility business to improve grid efficiency and stability.

2.7.1.2 Objective

The utility firm sought to accommodate the intermittent nature of solar electricity while maximizing the integration of solar energy into the system, reducing the reliance on fossil fuels, and maintaining grid stability.

2.7.1.3 Location

Southern California, USA

2.7.1.4 Data Sources

- Weather Stations: Solar irradiance, temperature, and wind speed data from a network of weather stations across the service area.
- Solar Farms: Historical solar energy generation records from solar farms connected to the grid.
- Grid Data: Historical energy demand and grid performance data.

2.7.1.5 Neural Network Architecture

A feed-forward neural network with the following architecture:

- Input Layer: Solar irradiance, temperature, and wind speed.
- Three Hidden Layers: Each with 128 neurons and ReLU activation functions.
- Output Layer: Solar energy generation prediction.

Data Split:

- Training Data: 70% of the dataset, spanning multiple years.
- Validation Data: 15% of the dataset for hyperparameter tuning.
- Testing Data: 15% of the dataset for model evaluation.

Feature Engineering:

- Time of day encoding using sine and cosine transformations.
- Lag features to capture historical trends.
- Normalization of input features to ensure consistent scales.

Model Training:

- Training Algorithm: Adam optimizer.
- Loss Function: Mean squared error (MSE).
- Learning Rate: 0.001.
- Epochs: 100.

Performance Metrics:

- MSE on testing data.
- Root mean squared error (RMSE) on testing data.
- Coefficient of determination (R-squared) on testing data.

Results:

- MSE: 5.2.
- RMSE: 2.28.
- R-squared (R^2): 0.92.

2.7.1.6 *Summary*

Based on solar irradiance, temperature, and wind speed, the neural network-based solar energy forecasting model showed great accuracy in predicting solar energy generation. This precise forecasting helped the utility company's service region integrate more renewable energy while also improving grid management and stability. The model's ability to reduce forecasting errors resulted in cost savings, optimized energy distribution, and a reduced carbon footprint.

2.7.2 Case Study 2: Fuzzy Logic Control for Maximum Power Point Tracking

Using fictitious data, this case study examines the use of fuzzy logic control in MPPT for a PV system as shown in Figure 2.5. For solar panels to produce as much electricity as possible under a variety of environmental circumstances, MPPT is crucial. A fuzzy logic-based MPPT system's design, implementation, and performance assessment is described in detail in the paper and validated by sample data.

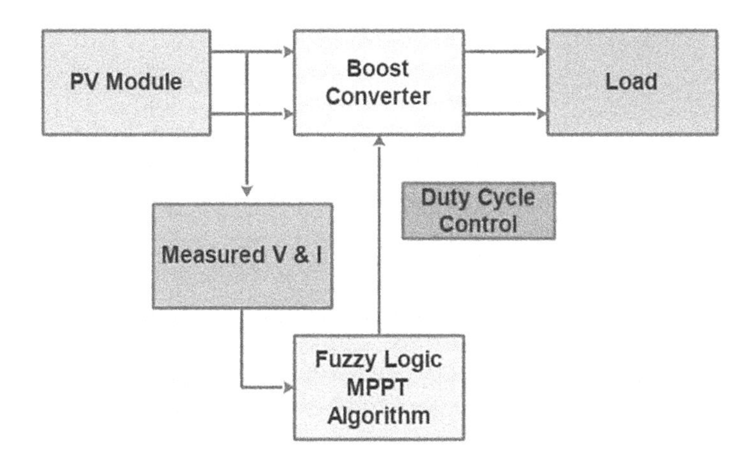

Figure 2.5 Design of fuzzy logic-based MPPT.

2.7.2.1 Introduction

Effective MPPT is essential for maximizing energy output in photovoltaic systems since solar energy is a valuable renewable resource. MPPT may be applied in real-world settings thanks to fuzzy logic control's adaptable and flexible approach.

2.7.2.2 Problem Statement

In this chapter, a fuzzy logic control system for MPPT is designed, put into practice, and evaluated using fictitious data.

2.7.2.3 Important Goals Include

- Constructing a fuzzy logic controller that can change to accommodate a changing environment.
- Integrating the controller into a PV system.
- Evaluating the system's performance in terms of stability and energy efficiency.

2.7.2.4 Methodology

2.7.2.4.1 Fuzzy Logic Controller Design

Solar radiation and temperature are used as input variables in the fuzzy logic controller's design, while voltage and current set points are used as output variables. Inputs are mapped to outputs using fuzzy rules. The controller makes use of language terms like "low," "medium," and "high."

2.7.2.4.2 Sample Data

A simplified illustration of potential data used for evaluation is given in Table 2.2.

2.7.2.4.3 Performance Evaluation

Using the example data given, the fuzzy logic-based MPPT system is assessed using metrics including

- Energy Efficiency
- Response Time
- Robustness
- Stability.

2.7.2.5 Results and Discussion

The fuzzy logic-based MPPT system's findings are shown in the case study utilizing fictional data. For instance, advantages in energy efficiency versus a fixed voltage system may be seen. The system's response to changing environmental conditions, robustness, and stability will be discussed.

2.7.2.6 Summary

Even using fictitious data, fuzzy logic control for MPPT in solar systems shows its ability to maximize energy output. The advantages in terms of energy economy, reactivity, and resilience are highlighted in this case study.

2.7.3 Case Study 3: Genetic Algorithm-Based PV Sizing

For residential applications, photovoltaic system efficiency sizing is essential for maximizing energy production and financial viability. In this case study, the use of GAs to size a residential PV system is examined. Strong optimization tools called GAs are renowned for their capacity to uncover almost ideal solutions in intricate, multidimensional problem environments [23]. To establish the ideal arrangement of panels, inverters, and batteries, GAs can consider a number of variables in the context of PV scaling, including location-specific solar irradiation, load profiles, and financial limitations.

Table 2.2 Potential Data for Evaluation of Performance of the System

Time (hours)	*Solar Radiation (W/m^2)*	*Temperature* $\left(C^0\right)$
8:00	300	25
9:00	600	30
10:00	800	35
11:00	700	30

Residential PV systems consist of solar panels, inverters, and potentially batteries to store excess energy. Finding the ideal mix of these elements to satisfy the household's energy requirements while staying within a given budget and taking regional climate variables into account is the difficult part. Traditional techniques for sizing frequently rely on human calculations or rule-of-thumb procedures, which may lead to systems that are either over- or undersized, resulting in subpar performance.

2.7.3.1 *Methodology*

For this case study, we gathered information given in Table 2.3 for a home site in Phoenix, Arizona, including daily load profiles (30 kWh/day) and solar irradiation (6.5 kWh/m2 per day). The only accessible area was 50 square metres, and the budget was restricted to $10,000. With a population mass of 100 and a running time of 50 generations, the GA was used to maximize energy production as considering the financial and physical restrictions.

2.7.3.2 *Summary*

Comparing the GA-optimized system to the rule-of-thumb and conventional optimization methodologies, the previous showed better performance in terms of energy production and cost-effectiveness. It was possible to find a solution that maximized energy output while considering budget and space restrictions by utilizing the capability of GAs to consider a wide variety of factors and constraints at the same time. This case study provides a more individualized and effective means of system design and demonstrates the practical advantages of employing genetic algorithms for PV scaling in residential applications.

The size of residential PV systems may be improved by using genetic algorithms, which enable a customized solution that maximizes energy production while taking financial and physical limits into account. This case study demonstrates how GAs have the potential to revolutionize solar system design and installation, paving the way for a more cost-efficient and sustainable future for residential solar energy.

Table 2.3 Comparison of Results

Metric	*GA-Optimizes*	*Rule-of-Thumb*	*Traditional Optimization*
Panel Count	8	10	9
Inverter Capacity (kW)	5	6	5.5
Battery Capacity (kWh)	20	24	22
Energy generation (kWh/year)	9125	8000	8750
Cost ($)	9800	11000	10500

2.8 CHALLENGES AND FUTURE DIRECTIONS

2.8.1 Current Challenges in Soft Computing Applications for Solar PV Systems

- **Data Quality and Availability:** Artificial neural networks (ANNs) and genetic algorithms are two examples of soft computing methods that strongly rely on high-quality data. It can be difficult to obtain precise, dependable, and comprehensive data for solar PV systems, such as solar irradiance, temperature, and historical energy output, particularly in areas with sparse monitoring infrastructure.
- **Data Size and Complexity:** Systems using solar PV produce an enormous amount of data. This huge data handling and processing can be computationally demanding and may call for cutting-edge methods for feature selection, dimensionality reduction, and effective data archiving and retrieval.
- **Model Generalization:** It is difficult to create soft computing models that generalize successfully across many locations and environmental situations. Due to differences in climate, shading, and other variables, models developed in one area may not function as well as they should in another.
- **Hardware and Computational Resources:** Soft computing methods may require a lot of computer power, particularly when working with huge datasets or challenging optimization issues. For certain academics and organizations, access to high-performance computer resources might be a limiting barrier.
- **Real-Time Operation:** Soft computing models must function in real-time for applications like grid integration and solar energy forecasting. It is quite difficult to achieve accurate forecasts with minimal latency.
- **Hybrid Models and Integration:** It might be challenging to combine various soft computing methods and incorporate them into current solar PV installations or grid infrastructure. It is a constant task to ensure smooth communication and collaboration among many components.
- **Robustness to Environmental Changes:** Soft computing models must adjust to shifting environmental factors, such as changing seasons and the weather. It might be difficult to maintain models' dependability and resilience over time.
- **Cost and Resource Optimization:** The design and operation of solar PV systems can be improved with the aid of soft computing, but finding the ideal balance between cost, resource utilization (such as battery storage), and energy output can be difficult.

- **Ethical and Security Concerns:** There are worries about data privacy, security, and ethical issues associated with AI/ML decision-making as soft computing models are incorporated further into energy infrastructure.

2.8.2 Future Research Avenues and Emerging Trends

Here are a few prospective future study directions and developing solar PV system trends:

- **Advanced PV Materials:** Novel PV materials being studied right now include tandem solar cells and perovskite solar cells. The efficacy and cost-effectiveness of solar panels might be considerably increased by these materials.
- **Bifacial Solar Panels:** Solar panels that can collect sunlight from both the front and back sides are becoming more and more common. The distribution and design of bifacial panels might be optimized in the future for increased energy production.
- **Flexible and Lightweight PV Technologies:** Solar panels that are flexible and light are becoming more popular because of the potential uses for wearable technology, portable electronics, and PV systems that are incorporated into buildings.
- **Machine Learning and AI for PV Systems:** PV system operation and maintenance are progressively being optimized using AI and machine learning. The precision of problem detection, predictive maintenance, and energy forecasting will likely be the key topics of future study.
- **PV System Durability and Recycling:** The need to guarantee the long-term durability of panels and create recycling procedures for PV modules that have reached the end of their useful lives will increase as there are more PV systems deployed.
- **PV System Resilience:** To assure the dependability of solar power generation in the face of climate change, research on the resilience of PV systems to extreme weather events, such as hurricanes, floods, and wildfires, is crucial.
- **Policy and Regulatory Research:** Ongoing research is being done on the laws, incentives, and policies that promote the expansion of the solar PV sector. This covers carbon pricing systems, feed-in tariffs, and net metering regulations.
- **Space-Based Solar Power:** A long-term idea is space-based solar power, which would send solar energy from space to Earth. Exploration of the viability, technological, and financial elements of space-based solar generation will continue.

2.9 CONCLUSION

In the realm of solar photovoltaic energy systems, soft computing approaches have become essential tools. The design, monitoring, and optimization of PV systems have been transformed by these flexible and data-driven approaches, which include artificial neural networks, genetic algorithms, fuzzy logic, and swarm intelligence. Soft computing offers more precise estimates of energy yield, optimal component size, and real-time control changes by successfully modeling complex and dynamic interactions inside PV systems. Additionally, by supporting grid stability through demand response and grid-friendly inverter designs, soft computing approaches make it easier to integrate PV installations into the larger energy environment. They also claim to improve environmental sustainability through improved load control and energy storage. Despite this, problems still exist, such as the need for varied and high-quality data, worries about model interpretability, and resource requirements for computing. Future research is anticipated to concentrate on improving existing techniques, overcoming these difficulties, and investigating novel applications including hybrid systems and space-based solar power as the area develops.

2.10 FUTURE WORK

The future work for this chapter can focus on several promising directions. Firstly, researchers may explore the integration of emerging soft computing techniques that have developed since the initial review, including advancements in genetic algorithms, swarm intelligence, and deep learning. Real-time applications, such as monitoring, control, and optimization of solar PV systems, could be a crucial area of investigation. Additionally, researchers could delve into fault detection and diagnostics using soft computing, aiming to enhance system reliability. Optimization and energy management strategies, as well as the development of data-driven decision support systems, may further contribute to the efficiency of solar PV systems. Exploring the integration of soft computing into cyber-physical systems, resilience analysis, and assessing environmental impacts could also be valuable research avenues. Lastly, collaborations across disciplines, including artificial intelligence, control systems, and renewable energy, may offer a multidisciplinary approach to advancing the field. Keeping abreast of the latest literature and technological developments will be essential for identifying emerging trends and fruitful areas for future exploration.

REFERENCES

[1] Paul A. Adedeji, Stephen A. Akinlabi, Nkosinathi Madushele, Obafemi O. Olatunji, "Beyond site suitability: Investigating temporal variability for utility-scale

solar-PV using soft computing techniques", *Renewable Energy Focus*, vol. 39, pp. 72–89, 2021. https://doi.org/10.1016/j.ref.2021.07.008
[2] B. A. Basit, Jin-Woo Jung, "Recent developments and future research recommendations of control strategies for wind and solar PV energy systems", *Energy Reports*, vol. 8, pp. 14318–14346, 2022. https://doi.org/10.1016/j.egyr.2022.10.395
[3] B. Aljafari, S. Devakirubakaran, C. Bharatiraja, P. K. Balachandran, T. Sudhakar Babu, "Power enhanced solar PV array configuration based on calcudoku puzzle pattern for partial shaded PV system", *Heliyon*, vol. 9, no. 5, 2023. https://doi.org/10.1016/j.heliyon.2023.e16041
[4] A. G. Olabi, M. Ali Abdelkareem, C. Semeraro, M. Al Radi, H. Rezk, O. Muhaisen, O. Adil Al-Isawi, E. T. Sayed, "Artificial neural networks applications in partially shaded PV systems", *Thermal Science and Engineering Progress*, vol. 37, 2023. https://doi.org/10.1016/j.tsep.2022.101612
[5] M. Balamurugan, S. Sahoo, S. Sukchai, "Application of soft computing methods for grid connected PV system: A technological and status review", *Renewable and Sustainable Energy Reviews*, vol. 75, 2016. https://doi.org/10.1016/j.rser.2016.11.210
[6] Z. Liu, et al., "Artificial intelligence powered large-scale renewable integrations in multi-energy systems for carbon neutrality transition: Challenges and future perspectives", *Energy and AI*, vol. 10, 2022.
[7] R. Shankar Archana, S. Singh, "Development of smart grid for the power sector in India", *Cleaner Energy Systems*, vol. 2, 2022.
[8] I. Rojek, D. Mikolajewski, A. Mrozinski, M. Macko, "Machine learning- and artificial intelligence-derived prediction for home smart energy systems with PV installation and battery energy storage", *Energies*, vol. 16, no. 18, 2023.
[9] C. Wang, M. H. Nehrir, "Power management of a stand-alone wind/photovoltaic/fuel cell energy system", *IEEE Transactions on Energy Conversion*, vol. 23, no. 3, pp. 957–967, 2008. https://doi.org/10.1109/TEC.2007.914200
[10] B. Armstrong, "FLC design for bounded separable functions with linear input-output relations as a special case", *IEEE Transactions on Fuzzy Systems*, vol. 4, no.1, pp. 72–79, 1996.
[11] C. Larbes, S. M. Aït Cheikh*, T. Obeidi, A. Zerguerras, "Genetic algorithms optimized fuzzy logic control for the maximum power point tracking in the photovoltaic system", *Renewable Energy*, vol. 34, no. 19, pp. 2093–2100–55, 2009.
[12] Tina Samavat, et al. "A comparative analysis of the Mamdani and Sugeno fuzzy inference systems for MPPT of an islanded PV system", *International Journal of Energy Research*, vol. 2023, 2023.
[13] P. Jayakumar & UN. *ESCAP Solar Energy: Resource Assessment Handbook.* [online]. Available at: https://hdl.handle.net/20.500.12870/5255
[14] S. Salcedo-Sanz, C. Casanova-Mateo, A. Pastor-Sánchez, D. Gallo-Marazuela, A. Labajo-Salazar, A. Portilla-Figueras, "Direct solar radiation prediction based on soft-computing algorithms including novel predictive atmospheric variables", in: *Intelligent Data Engineering and Automated Learning – IDEAL 2013. IDEAL 2013. Lecture Notes in Computer Science*, vol. 8206. Springer, Berlin, Heidelberg. https://doi.org/10.1007/978-3-642-41278-3_39
[15] T. Verma, A. P. S. Tiwana, C. C. Reddy, V. Arora, P. Devanand, "Data analysis to generate models based on neural network and regression for solar power generation forecasting", *2016 7th International Conference on Intelligent*

Systems, Modelling and Simulation (ISMS), Bangkok, Thailand, pp. 97–100, 2016. https://doi.org/10.1109/ISMS.2016.65

[16] M. Ilbeigi, M. Ghomeishi, A. Dehghanbanadaki, "Prediction and optimization of energy consumption in an office building using artificial neural network and a genetic algorithm", *Sustainable Cities and Society*, vol. 61, 2020.

[17] S. Korjani, A. Serpi, A. Damiano, "A genetic algorithm approach for sizing integrated PV-BESS systems for prosumers", *2020 2nd IEEE International Conference on Industrial Electronics for Sustainable Energy Systems (IESES)*, Cagliari, Italy, pp. 151–156, 2020. https://doi.org/10.1109/IESES45645.2020.9210700

[18] T. Klinsuwan, W. Ratiphaphongthon, R. Wangkeeree, R. Wangkeeree, C. Sirisamphanwong, "Evaluation of machine learning algorithms for supervised anomaly detection and comparison between static and dynamic thresholds in photovoltaic systems" *Energies*, vol. 16, no. 4, p. 1947, Feb. 2023. https://doi.org/10.3390/en16041947

[19] M. A. Nasab, et al., "Uncertainty compensation with coordinated control of EVs and DER systems in smart grids", *Solar Energy*, vol. 263, 2023.

[20] T. Atasoy, H. E. Akinc, O. Ercin, "An analysis on smart grid applications and grid integration of renewable energy systems in smart cities", *2015 International Conference on Renewable Energy Research and Applications (ICRERA)*, Palermo, Italy, pp. 547–550, 2015.

[21] M. R. Maghami, J. Pasupuleti, C. M. Ling, "Comparative analysis of smart grid solar integration in urban and rural networks", *Smart Cities*, vol. 6, no. 5, pp. 2593–2618, 2023.

[22] T. A. Jumani, M. W. Mustafa, A. S. Alghamdi, M. M. Rasid, A. Alamgir, A. B. Awan, "Swarm intelligence-based optimization techniques for dynamic response and power quality enhancement of AC microgrids: A comprehensive review", *IEEE Access*, vol. 8, pp. 75986–76001, 2020. https://doi.org/10.1109/ACCESS.2020.2989133

[23] S. I. Sulaiman, T. K. A. Rahman, I. Musirin, S. Shaari, "Sizing grid-connected photovoltaic system using genetic algorithm", *2011 IEEE Symposium on Industrial Electronics and Applications*, ISIEA 2011, pp. 505–509, 2011. https://doi.org/10.1109/ISIEA.2011.6108763

Chapter 3

Parameter Identification for Three-Diode Model Using the Tiki Taka Algorithm

Moncef El Marghichi, Azeddine Loulijat, Abdelilah Hilali, and Abdelhak Essounaini

3.1 INTRODUCTION: BACKGROUND AND DRIVING FORCES

Rcliable, sustainable energy is crucial for humanity's well-being. The depletion of traditional sources calls for a transition to renewables. Solar and wind power advancements have enhanced energy production. Solar PV technology finds various applications, including water desalination and cooling/heating systems [1].

Accuracy modelling and estimation of solar cells are crucial for understanding their behavior. Different approaches are used to develop photovoltaic models. The single diode model (SDM) is simple and efficient. The double diode model (DDM) improves accuracy by including losses in both the SDM and the space charge region. For even higher precision, the three-diode model (TDM) incorporates losses in the defect region [2].

Accurate estimation of PV cell model parameters is crucial for optimizing and implementing PV systems. The challenges presented by the nonconvex and nonlinear characteristics of the photovoltaic model have prompted researchers to address them through the development of three distinct approaches for accurate parameter estimation. These methods include analytic, metaheuristic, and deterministic techniques [3].

Analytic methods formulate equations based on specific data points, such as open-circuit and short-circuit points, to handle parameters in the PV model. While these approaches offer speed and convenience, their effectiveness is heavily contingent upon the availability of precise data supplied by manufacturers, which may compromise accuracy. Additionally, they might be vulnerable to the influence of photovoltaic degradation over time [4, 5].

Deterministic methods necessitate a substantial quantity of measurements and utilize a loss function to assess the difference between estimated and observed data points. Such approaches tend to lead to locally optimal solutions by leveraging gradient information [6–8].

Metaheuristic techniques are popular for precise parameter extraction in PV models due to their nonlinearity. Various approaches have been developed to achieve high precision. For instance, the method introduced in [9]

DOI: 10.1201/9781003462460-3

merges cuckoo search algorithm (CSA) and the grey wolf optimizer (GWO) to balance exploitation and exploration. The MLBSA accurately estimates PV parameters [10], while the GBO (gradient-based optimizer) proves effective in parameter estimation for different PV models [11]. Additional methods like the sunflower optimization algorithm, modified JAYA algorithm, WOA (whale optimization algorithm), CLJAYA (comprehensive learning JAYA algorithm), WSO, and NGO (northern goshawk optimization) have also been used for parameter extraction [12–17].

Hybrid strategies have been investigated for obtaining parameters of the photovoltaic model [18–21]. However, it is important to note that no single metaheuristic optimization technique can universally solve all optimization problems, as stated by the no free lunch (NFL) theorem [22]. This theorem emphasizes the need to adjust available techniques for specific problem domains.

The study of determining solar cell parameters and utilizing metaheuristics has identified specific limitations. These encompass non-adaptive weight metrics, sluggish calculation speed, vulnerability to traps of local optimum, and the need to minimize the root mean square error (RMSE) value.

To address these limitations, this chapter presents a novel search mechanism for parameter estimation in PV cells. The proposed approach utilizes the tiki taka algorithm (TTA) algorithm, inspired by the behavior of dung beetles, to efficiently optimize the parameters of the three-diode model (TDM).

TTA, inspired by football, utilizes a diverse set of key players to optimize problems. These players represent the best solutions and are updated based on fitness. By balancing exploration and exploitation, the algorithm maintains solution diversity while improving fitness. It emphasizes coordination and movement among key players to optimize the problem. The tiki taka algorithm offers an efficient and effective approach to exploration and optimization.

To evaluate the algorithm's performance, real solar modules (STP6–120/36 and Photowatt-PWP201) were used. Comparative analysis against six robust strategies confirmed the algorithm's resilience, effectiveness, and speed. In summary, the study contributes the following significant findings:

- This chapter introduces a novel TTA approach for estimating parameters for TDM.
- A comparative assessment of the TTA algorithm is conducted versus six robust algorithms, using the Photowatt-PWP201 and STP6–120/36 PV modules.
- The validation of the proposed algorithm's efficacy is conducted by assessing absolute power and current errors.
- Additionally, P-V and I-V curves are simulated using the extracted parameter to visually demonstrate the efficiency of TTA algorithm.

The chapter is structured as follows: in Section 3.2, the TDM PV model and relevant equations are introduced. Section 3.3 offers an overview of the TTA algorithm, followed by a description of the implementation setup in Section 3.4. The results are presented in Section 3.5, and Section 3.6 serves as the concluding part summarizing the findings.

3.2 TDM MODELLING

In this section, we explore the mathematical formulations pertaining to the TDM model for both cells and modules.

The photovoltaic generator incorporates a current source (I_{ph}) in parallel with three diodes and a resistor (R_{sh}), connected in series with another resistor (R_s) (see Figure 3.1). The current I_{ph} is distributed among the parallel diodes and resistor, yielding the generated current by the photovoltaic unit, expressed as follows:

$$I_{out} = I_{ph} - \sum_{j=1}^{m} I_{oj} - \frac{V + I_{out} R_s}{R_{sh}} \tag{3.1}$$

Here, the symbol m represents the number of parallel diodes, with m equal to 3. The output voltage is denoted by V, and I_{oj} is the current in diode "j" expressed as:

$$I_{oj} = I_{stj} \left(e^{q\left(\frac{V + I_{out} R_s}{njKT}\right)} - 1 \right) \tag{3.2}$$

I_{stj} symbolizes the saturation current, where q is the charge of one electron (1.602 e-19 C). Furthermore, K denotes the Boltzmann constant, n_j represents the ideality factor of the diode, and T is the temperature in Kelvin.

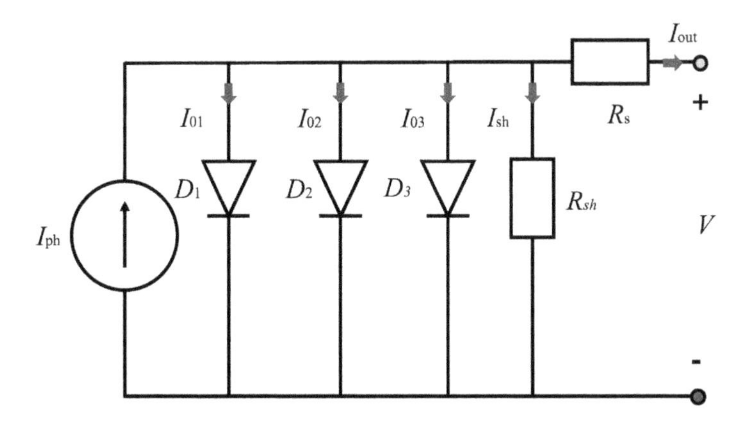

Figure 3.1 TDM model.

Combining Equations (3.1) and (3.2) yields the following expression:

$$I_{out} = I_{ph} - \sum_{j=1}^{p} I_{stj}\left(e^{q\left(\frac{V+I_{out}R_s}{njKT}\right)} - 1\right) - \frac{V + I_{out}R_s}{R_{sh}} \tag{3.3}$$

3.2.1 Photovoltaic Module Model

The current (I_{out}) of a photovoltaic (PV) module, with $N_s \times N_p$ solar cells arranged in series and/or parallel within a three-diode model (TDM) based PV module, is articulated as:

$$I_{out} = I_{ph} - \sum_{j=1}^{p} I_{stj}\left(e^{q\left(\frac{V/N_s+(I_{out}R_s)/N_p}{n_jKT}\right)} - 1\right) - \frac{V/N_s + (I_{out}R_s)/N_p}{R_{sh}} \tag{3.4}$$

3.2.2 Loss Function

The study aims to minimize the gap between simulated and measured current in solar cells. The widely adopted approach for this is using root mean square error as the loss function. The TTA algorithm is employed to estimate photovoltaic model parameters by minimizing RMSE. The loss function is defined by the difference between estimated and measured current:

$$F_{cost} = \sqrt{\frac{1}{W}\left(\sum_{j=1}^{W} \left|I_{es}(j) - I_{mes}(j)\right|^2\right)} \tag{3.5}$$

In the given equation, I_{es} and I_{mes} are the estimated and the experimental current. W represents the total number of data points.

3.3 TIKI TAKA ALGORITHM

TTA is a football-inspired metaheuristic that mimics the renowned playing style [23]. It utilizes triangular formations, short passing, and ball possession for game control. Key players strategically move within these formations to score goals. Candidates represent player positions, with influential leaders guiding the ball's position. TTA integrates tiki taka tactics for informed decision-making. TTA consists of three distinct phases: initialization, update ball position, and update player position.

3.3.1 Initialization

A football team with n players in d dimensions is considered, with bounded initial positions (LB and UB). Key players (nk) represent a minimum of three players or 10% of the total. Ball and player positions are stored in matrices B and P, respectively. Initial player positions are randomly generated using Equation (3.6).

$$p^{t+1}{}_i = rand * (UB - LB) + LB \tag{3.6}$$

Key players (nk) update their positions in each iteration, stored in the key players archive (h). Initially, B is the same as P. Equations (3.7) and (3.8) represent the matrix updates.

$$B = \begin{pmatrix} b^t{}_{1,1} & \cdots\cdot & b^t{}_{1,d} \\ . & \cdots\cdot & . \\ b^t{}_{N,1} & \cdots\cdot & b^t{}_{n,d} \end{pmatrix} \tag{3.7}$$

$$P = \begin{pmatrix} p^t{}_{1,1} & \cdots\cdot & p^t{}_{1,d} \\ . & \cdots\cdot & . \\ p^t{}_{N,1} & \cdots\cdot & p^t{}_{n,d} \end{pmatrix} \tag{3.8}$$

3.3.2 Upgrade Ball Position

Tiki taka employs short passing to update the ball position. Unsuccessful passes (10–30% with probability parameter u) are considered. Ball position is updated using:

$$b_i^{t+1} = \begin{Bmatrix} rand(b_i^t - b_i^{t+1}) + b_i^t, & r_p > \varphi \\ -(c_1 + rand) + b_i + (b_i - b_{i+1}), & r_p \le \varphi \end{Bmatrix} \tag{3.9}$$

Successful and unsuccessful passes are determined using a random number r_p. The ball position is updated based on these passes, with a reflection coefficient c_1. If the next ball position is equal to the last position, it is set to the initial position.

3.3.3 Update Key Players

Equation (3.10) updates the position of the i_{th} player based on the ball and key player positions in TTA.

$$p_i^{t+1} = p_i^t + rand * c_2 * (b_i^t - p_i^t) + rand * c_3 * (h - p_i^t) \tag{3.10}$$

The new position of the i_{th} player in TTA is determined by balancing between the ball and the key player, with c_2 (ranging from 1.0 to 2.5) and c_3 (ranging from 0.5 to 1.5) as the coefficients. The global best is represented by h.

Figure 3.2 illustrates the TTA pseudocode.

3.3.4 Proposed Algorithm for PV Parameter Estimation

The proposed algorithm in Figure 3.3 extracts the solar model parameter. It initializes the solar PV model, reads current and voltage measurements, and utilizes the TTA algorithm to find the optimal candidate (pbest) that minimizes the cost function Equation (3.5). TTA then outputs the best solution, the TDM model.

3.4 SETUP

The TTA algorithm (Figure 3.3) is employed for evaluating solar PV model parameters with STP6–120/36 and PWP201 PV modules [24]. Comparative analysis includes EO (equilibrium optimizer), GWO, RUN (runge kutta optimizer), SMA (slime mould algorithm), WOA, and GBO [25–30]. STP6–120/36 is a monocrystalline module, and PWP201 is polycrystalline, each with 36 series cells (Table 3.1). Parameters obtained using the TTA algorithm are detailed in Table 3.2 and Table 3.3.

3.5 RESULTS AND DISCUSSION

The goal is to determine the nine parameters (Ist1, Ist2, Ist3, Iph, Rsh, Rs, n1, n2, and n3) for the TDM PV modules STP6–120/36 and PWP201. Table 3.3

1- **begin**
2- Initialize the population number N_{pop},and the maximum iteration T_{max}.
3- Initialize the probability list parameter, and ball reflection magnitude c1.
4- Initialize the player to ball and player to current best one coefficients c2 and c3 .
5- Set initialize ball and player positions.
6- **while** (t < Tmax)
7- **for** each ball position b_i^t **do**
8- **if** (rand(0,1) > φ)
9- $b_i^{t+1} = b_i^t + rand(0,1).(b_i^t - b_i^{t+1})$
10- **else**
11- $b_i^{t+1} = b_i^t - (c1 + rand(0,1)).(b_i^t - b_i^{t+1})$
12- **endif**
13- **end for**
14- **for** each ball position p_i^t **do**
15- $p_i^{t+1} = p_i^t - (c2 * rand(0,1)).(b_i^t - p_i^t) + rand(0,1)*c3*(h - p_i^t)$
16- **end for**
17- t=t+1
18- break while loop if (fitness evaluation ≥max_fitness_evaluation)
19- **end while**
20- return the global best solution p_{best}
21- **end**

Figure 3.2 TTA pseudocode.

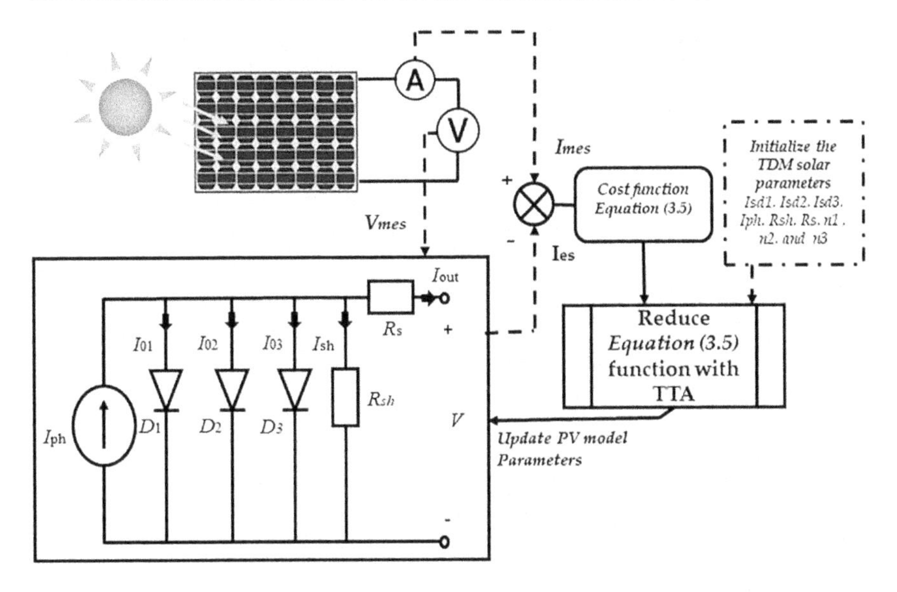

Figure 3.3 Framework used to estimate the TDM model parameters.

Table 3.1 PV Model

PV Type	*Temperature (°C)*	*(Np × Ns) Cells*
Photowatt-PWP201	45	1 × 36
STP6–120/36	55	1 × 36

Table 3.2 TTA Parameters

PV Type	*Population Number (N)*	*Number of Iterations T_{max}*	*Number of Decision Variables*
Photowatt-PWP201	50	1500	9
PV type	50	1500	9

Table 3.3 Limits of TDM Model

	STP6–120/36		*PWP201*	
Parameters	*Ub*	*Lb*	*Ub*	*Lb*
I_{st1}, I_{st2}, I_{st3}	50	0	50	0
I_{ph}	8	0	2	0
R_{sh}	1500	0	2000	0
R_s	0.36	0	2	0
n_3, n_2, n_1,	50	1	50	1

shows the parameter boundaries. Performance analysis is conducted using TTA and other algorithms, with P-V and I-V characteristics presented in Figure 3.4 and Figure 3.5. Convergence curves and absolute current error are shown in Figure 3.6, Figure 3.7, Figure 3.8, and Figure 3.9. Extracted parameters by TTA are listed in Table 3.4.

To evaluate method accuracy, we use three metrics: mean square error (MSE), root mean square error (RMSE), and normalized RMSE (NRMSE):

$$\text{MSE} = \frac{1}{A}\sum_{k=1}^{A}\left(I_{es}(k) - \text{I}_{mes}(k)\right)^2 \tag{9.11}$$

$$\text{NRMSE} = \frac{RMSE}{I_{es,\max} - I_{es,\min}} \tag{9.12}$$

$$\text{RMSE} = \sqrt{\frac{1}{A}\sum_{k=1}^{A}\left(I_{es}(k) - I_{mes}(k)\right)^2} \tag{9.13}$$

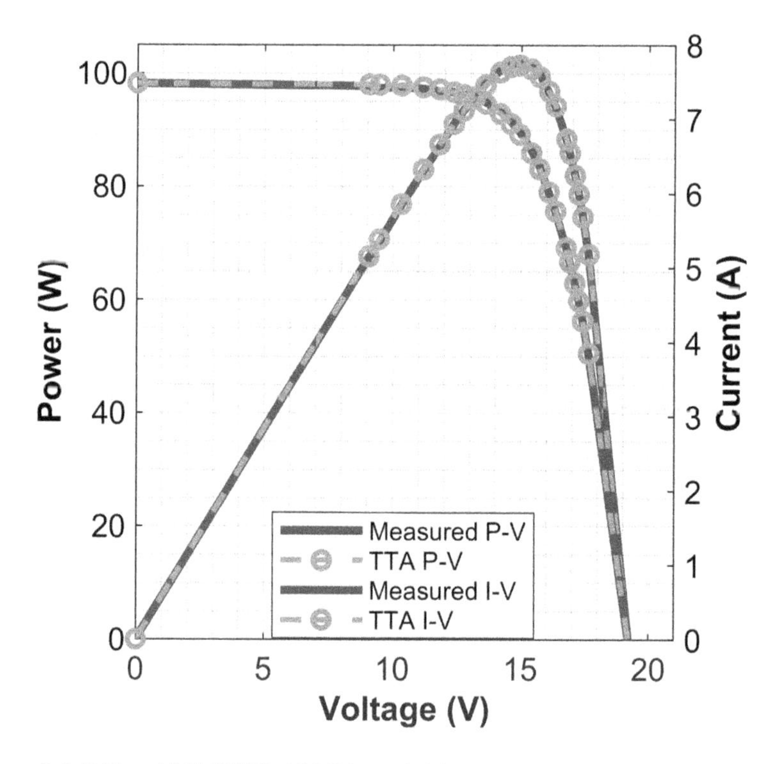

Figure 3.4 P-V and I-V (STP6–120/36 module).

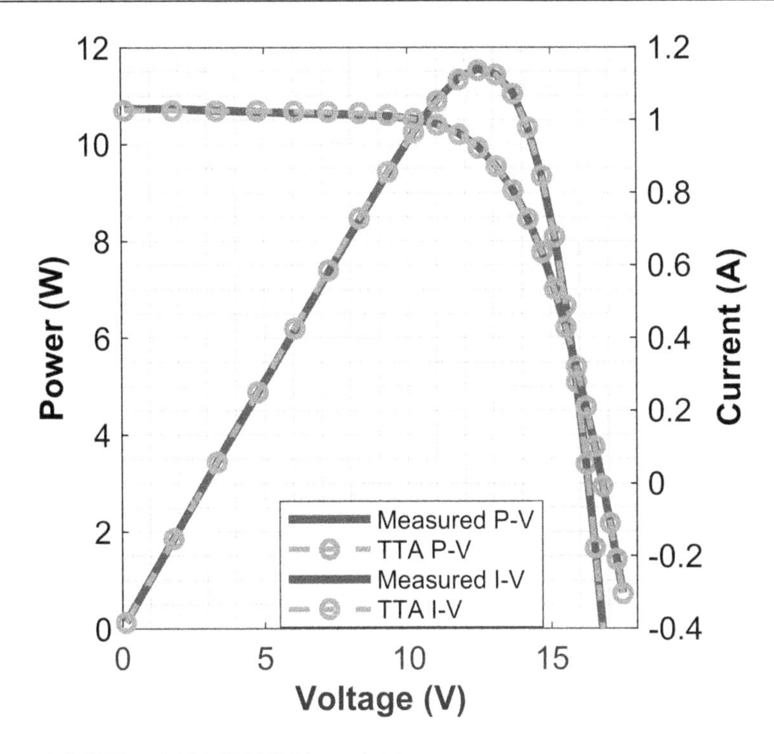

Figure 3.5 P-V and I-V (PWP201 module).

A represents the overall count of data points, where I_{es} and I_{mes} indicate the recorded and predicted output current values, respectively.

Table 3.5 presents a summary of the computed metrics for the different techniques employed in assessing PV solar parameters for the STP6 and PWP201 modules. The effectiveness of the algorithms is assessed through metrics including MSE, RMSE, and NRMSE.

TTA demonstrates outstanding performance compared to other algorithms, as indicated by the numerical results in Table 3.5. Focusing on the STP6–120/36 PV type, TTA achieves an RMSE of 0.0214 A, outperforming alternative methods such as EO (RMSE = 0.0511 A), GWO (RMSE = 0.0575 A), RUN (RMSE = 0.0453 A), SMA (RMSE = 0.0163 A), WOA (RMSE = 0.0250 A), and GBO (RMSE = 0.0206 A). This result demonstrates the superior accuracy of TTA in estimating the output current for the STP6–120/36 module. Additionally, TTA exhibits a significantly lower NRMSE value of 0.0029 and an MSE value of 4.5790e-04, further highlighting its effectiveness in predicting the output current.

Turning to the PWP201 PV type, TTA continues to outshine the other algorithms. It achieves an impressively low RMSE of 0.0026 A, while alternative methods like EO (RMSE = 0.0067 A), GWO (RMSE = 0.0037 A), RUN (RMSE = 0.0048 A), SMA (RMSE = 0.0040 A), WOA (RMSE = 0.0029 A),

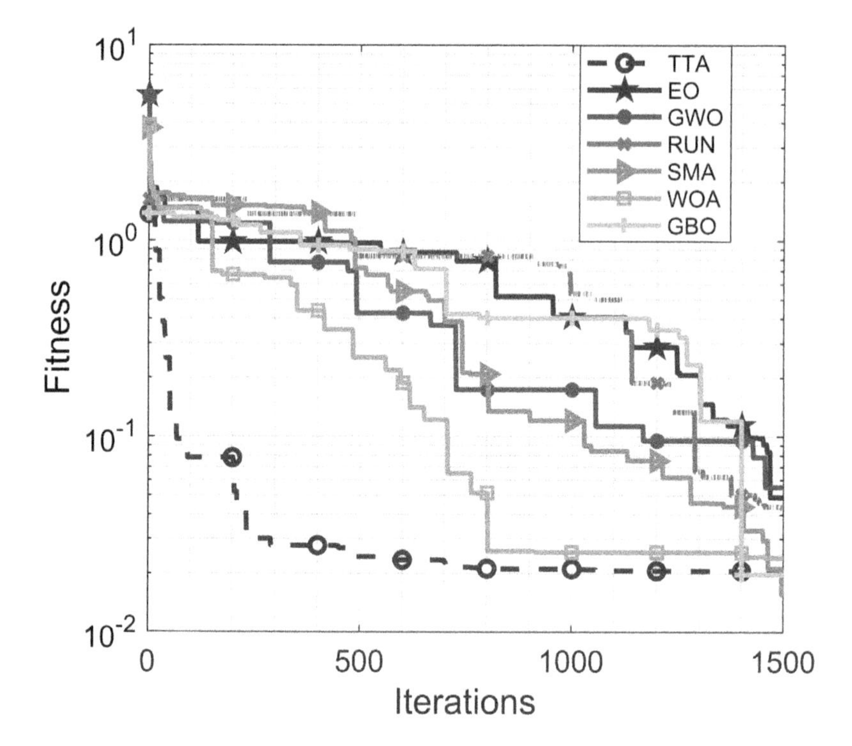

Figure 3.6 Fitness (STP6–120/36 module).

and GBO (RMSE = 0.0087 A) exhibit higher values. The exceptional accuracy of TTA in estimating the output current for the PWP201 module is further demonstrated by its NRMSE value of 0.0020 and an extremely low MSE value of 7.0025e-06. These results emphasize the superiority of TTA over the other algorithms in predicting the output current for the PWP201 PV type.

Table 3.6 presents a comprehensive analysis of different algorithms used for various solar PV panel types. The evaluation is based on several performance metrics, including max, min, mean, and power errors (as defined in Equation [3.14]), measured in mA and mW. A lower error value signifies superior algorithm performance, with decreased deviation from the actual values.

$$P\_err = \frac{1}{A}\sqrt{\sum_{k=1}^{A}\left|P_{mes}(k) - P_{es}(k)\right|}, \quad (3.14)$$

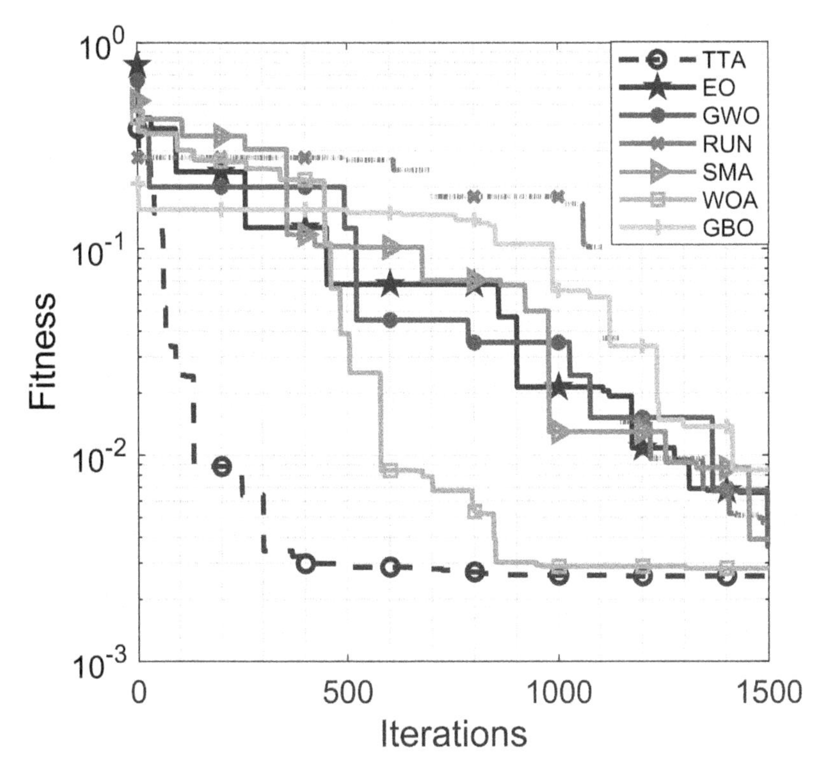

Figure 3.7 Fitness (PWP201 module).

P_{es} and P_{mes} are the evaluated and measured power, A denoting the number of points.

Table 3.6 provides a comprehensive comparison of algorithm performance, where the TTA stands out for its impressive results. Specifically, for the STP6–120/36 PV type, the TTA achieves a maximum error of 45.534174 mA, surpassing EO (50.059053 mA), GWO (80.192316 mA), WOA (59.808967 mA), and RUN (75.186037 mA), while performing slightly lower than SMA (37.909155 mA) and GBO (35.632347 mA). Additionally, the TTA excels in minimizing the minimum error, showcasing a value of 0.317686 mA, surpassing the other algorithms. Furthermore, the TTA demonstrates competitive performance in terms of mean error, recording a value of 17.655764 mA, and power error, measuring 245.7386 W (comparable to SMA and GBO). These results indicate the TTA's commendable accuracy in estimating the output current and power for the STP6–120/36 module.

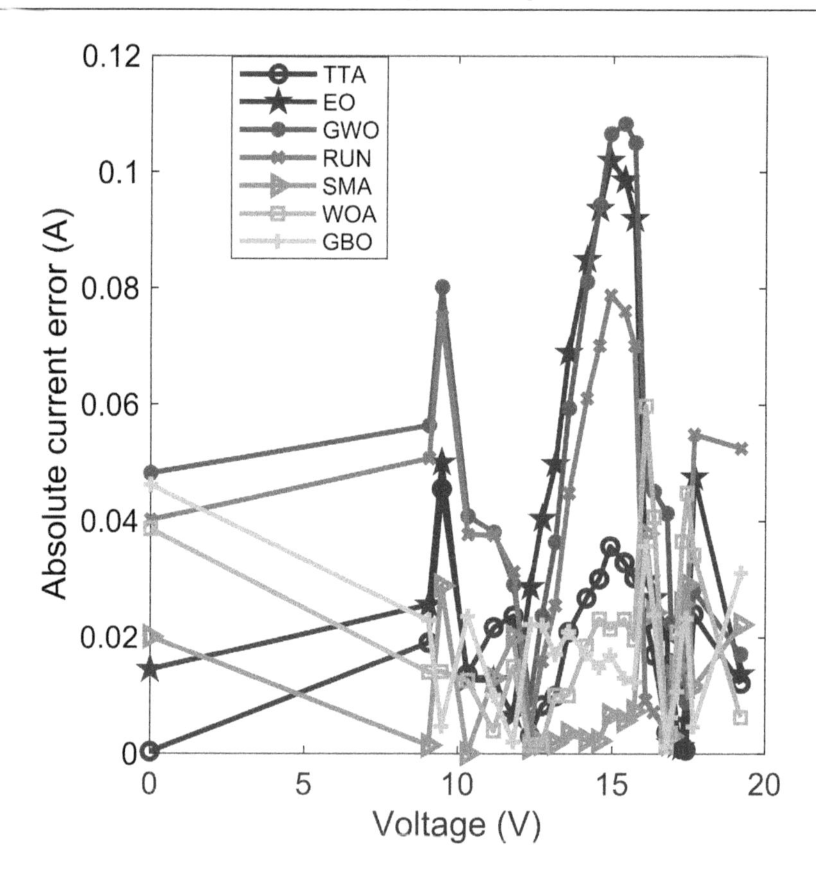

Figure 3.8 Absolute current in TDM model (STP6–120/36).

Likewise, when considering the PWP201 PV type, the TTA continues to showcase excellent performance. It achieves a maximum error of 5.428451 mA, surpassing the errors obtained by EO (13.855075 mA), RUN (7.116245 mA), and SMA (8.340543 mA), albeit slightly higher than GWO (4.176346 mA), WOA (4.315851 mA), and GBO (1.410660 mA). Notably, the TTA stands out by demonstrating the lowest mean error of 2.020256 mA and power error of 16.8483 W, further highlighting its superior accuracy in estimating the output current and power for the PWP201 module.

In conclusion, the TTA displays exceptional performance when compared to other algorithms for both the STP6–120/36 and PWP201 PV types. For the STP6–120/36 module, TTA achieves competitive results, outperforming several algorithms in terms of maximum and minimum errors, while exhibiting a commendable mean error and power error. Similarly, for the PWP201 module, TTA excels by demonstrating lower

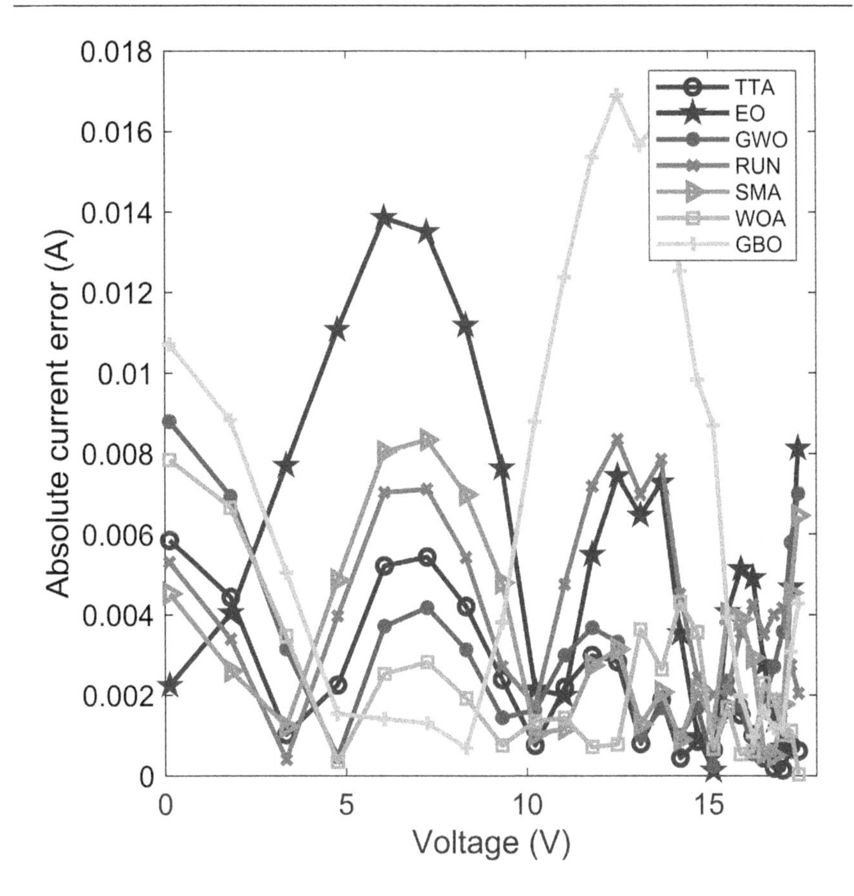

Figure 3.9 Absolute current in TDM model (PWP201).

Table 3.4 TTA-Derived Parameters for the TDM Model

PV Type	I_{st1}, I_{st2} I_{st3} *(μA)*	I_{ph} *(A)*	R_{sh} *(Ω)*	R_s *(Ω)*	n_1, n_2, n_3
STP6–120/36	6.506; 13.88; 9.1593	7.4797	1091.352	0.00403	1.449; 43.94; 45.890
PWP201	5.62; 25.365; 49.233	1.0265	86.611	0.03201	1.4588; 35.202; 35.586

maximum errors and achieving the lowest mean error and power error among the algorithms considered. These results highlight the TTA's superior accuracy in estimating the output current and power for both PV types, emphasizing its effectiveness in PV system modeling and performance evaluation.

Table 3.5 Performance Metrics for the TDM Model Predictions

PV Type	*Methods*	*RMSE (A)*	*NRMSE*	*MSE*
STP6–120/36	TTA	0.0214	0.0029	4.5790e-04
	EO	0.0511	0.0068	0.0026
	GWO	0.0575	0.0077	0.0033
	RUN	0.0453	0.0061	0.0020
	SMA	0.0163	0.0022	2.6620e-04
	WOA	0.0250	0.0033	6.2478e-04
	GBO	0.0206	0.0027	4.2264e-04
Photowatt-PWP201	TTA	0.0026	0.0020	7.0025e-06
	EO	0.0067	0.0050	4.5143e-05
	GWO	0.0037	0.0028	1.3509e-05
	RUN	0.0048	0.0036	2.2909e-05
	SMA	0.0040	0.0030	1.5879e-05
	WOA	0.0029	0.0022	8.3284e-06
	GBO	0.0087	0.0065	7.5449e-05

Table 3.6 Absolute Maximum, Minimum, Mean, and Power Errors of the Algorithms

PV Type	*Methods*	*Max (mA)*	*Min (mA)*	*Mean (mA)*	*Power Error (W)*
STP6–120/36	TTA	45.534174	0.317686	17.655764	245.7386
	EO	50.059053	1.182007	39.017020	551.2387
	GWO	80.192316	8.892480	48.692010	660.7151
	RUN	75.186037	3.770799	37.932255	511.9202
	SMA	37.909155	0.054444	12.029364	171.0401
	WOA	59.808967	1.296168	19.813000	279.4468
	GBO	35.632347	1.220461	17.659928	230.1119
Photowatt-PWP201	TTA	5.428451	0.145858	2.020256	16.8483
	EO	13.855075	0.132453	5.537981	56.5005
	GWO	4.176346	0.238957	2.983887	30.5688
	RUN	7.116245	0.104693	4.302360	48.7238
	SMA	8.340543	0.501482	3.272569	34.3732
	WOA	4.315851	0.035794	2.185173	19.6712
	GBO	1.410660	0.692580	6.723827	76.2813

3.6 CONCLUSION

This study introduces the TTA algorithm, inspired by football, for parameter extraction in the three-diode model. TTA utilizes a diverse set of key players to optimize solutions, updating them based on fitness to maintain diversity and improve performance. The TTA algorithm efficiently explores and finds optimal solutions, minimizing unnecessary exploration.

The TTA algorithm's effectiveness is exemplified in parameter extraction for the three-diode model (TDM) within the STM6–40/36 and

Photowatt-PWP201 PV modules. TTA outperforms state-of-the-art algorithms with remarkably low RMSE values of 0.0214 A and 0.0026 A and achieves the lowest power errors of 245.7386 mW and 16.8483 mW for the respective modules. These findings underscore the superior accuracy of the TTA algorithm in estimating parameters for photovoltaic models.

REFERENCES

[1] M. Mostafa, H. M. Abdullah, and M. A. Mohamed, "Modeling and experimental investigation of solar stills for enhancing water desalination process", *IEEE Access*, vol. 8, 2020, pp. 219457–219472.
[2] M. A. Soliman, H. M. Hasanien, and A. Alkuhayli, "Marine predators algorithm for parameters identification of triple-diode photovoltaic models", *IEEE Access*, **vol.** 8, 2020, pp. 155832–155842.
[3] E. I. Batzelis and S. A. Papathanassiou, "A method for the analytical extraction of the single-diode PV model parameters", *IEEE Transactions on Sustainable Energy*, **vol.** 7, no. 2, 2015, pp. 504–512.
[4] P. Changmai, S. K. Nayak, and S. K. Metya, "Estimation of PV module parameters from the manufacturer's datasheet for MPP estimation", *IET Renewable Power Generation*, **vol. 14**, no. 11, 2020, pp. 1988–1996.
[5] Y.-C. Huang, C.-M. Huang, S.-J. Chen, and S.-P. Yang, "Optimization of module parameters for PV power estimation using a hybrid algorithm", *IEEE Transactions on Sustainable Energy*, **vol. 11**, no. 4, 2019, pp. 2210–2219.
[6] D. H. Muhsen, A. B. Ghazali, T. Khatib, and I. A. Abed, "Parameters extraction of double diode photovoltaic module's model based on hybrid evolutionary algorithm", *Energy Conversion and Management*, **vol. 105**, 2015 pp. 552–561.
[7] S. Gao, K. Wang, S. Tao, T. Jin, H. Dai, and J. Cheng, "A state-of-the-art differential evolution algorithm for parameter estimation of solar photovoltaic models", *Energy Conversion and Management*, **vol. 230**, 2021, p. 113784.
[8] D. Saadaoui, M. Elyaqouti, K. Assalaou, and S. Lidaighbi, "Parameters optimization of solar PV cell/module using genetic algorithm based on non-uniform mutation", *Energy Conversion and Management: X*, **vol. 12**, 2021, p. 100129.
[9] W. Long, S. Cai, J. Jiao, M. Xu, and T. Wu, "A new hybrid algorithm based on grey wolf optimizer and cuckoo search for parameter extraction of solar photovoltaic models", *Energy Conversion and Management*, **vol. 203**, Jan. 2020, p. 112243.
[10] K. Yu, J. J. Liang, B. Y. Qu, Z. Cheng, and H. Wang, "Multiple learning backtracking search algorithm for estimating parameters of photovoltaic models", *Applied Energy*, **vol. 226**, Sep. 2018, pp. 408–422.
[11] A. A. Ismaeel, E. H. Houssein, D. Oliva, and M. Said, "Gradient-based optimizer for parameter extraction in photovoltaic models", *IEEE Access*, **vol. 9**, 2021, pp. 13403–13416.
[12] M. H. Qais, H. M. Hasanien, and S. Alghuwainem, "Identification of electrical parameters for three-diode photovoltaic model using analytical and sunflower optimization algorithm", *Applied Energy*, **vol. 250**, 2019, pp. 109–117.
[13] T. V. Luu and N. S. Nguyen, "Parameters extraction of solar cells using modified JAYA algorithm", *Optik*, **vol. 203**, 2020, p. 164034.

[14] O. S. Elazab, H. M. Hasanien, M. A. Elgendy, and A. M. Abdeen, "Parameters estimation of single-and multiple-diode photovoltaic model using whale optimisation algorithm", *IET Renewable Power Generation*, **vol. 12**, no. 15, 2018, pp. 1755–1761.
[15] Y. Zhang, M. Ma, and Z. Jin, "Comprehensive learning Jaya algorithm for parameter extraction of photovoltaic models", *Energy*, **vol. 211**, 2020, p. 118644.
[16] T. S. Ayyarao and P. P. Kumar, "Parameter estimation of solar PV models with a new proposed war strategy optimization algorithm", *International Journal of Energy Research*, **vol. 46**, no. 6, 2022, pp. 7215–7238.
[17] M. A. El-Dabah, R. A. El-Sehiemy, H. M. Hasanien, and B. Saad, "Photovoltaic model parameters identification using Northern Goshawk Optimization algorithm", *Energy*, **vol. 262**, 2023, p. 125522.
[18] W. Long, S. Cai, J. Jiao, M. Xu, and T. Wu, "A new hybrid algorithm based on grey wolf optimizer and cuckoo search for parameter extraction of solar photovoltaic models", *Energy Conversion and Management*, **vol. 203**, 2020, p. 112243.
[19] S. Li, W. Gong, L. Wang, X. Yan, and C. Hu, "A hybrid adaptive teaching – learning-based optimization and differential evolution for parameter identification of photovoltaic models", *Energy Conversion and Management*, **vol. 225**, 2020, p. 113474.
[20] S. Wang, Y. Yu, and W. Hu, "Static and dynamic solar photovoltaic models' parameters estimation using hybrid Rao optimization algorithm", *Journal of Cleaner Production*, **vol. 315**, 2021, p. 128080.
[21] J. P. Ram, T. S. Babu, T. Dragicevic, and N. Rajasekar, "A new hybrid bee pollinator flower pollination algorithm for solar PV parameter estimation", *Energy Conversion and Management*, **vol. 135**, 2017, pp. 463–476.
[22] S. P. Adam, S.-A. N. Alexandropoulos, P. M. Pardalos, and M. N. Vrahatis, "No free lunch theorem: A review", in *Approximation and Optimization: Algorithms, Complexity and Applications*, pp. 57–82, 2019. https://link.springer.com/chapter/10.1007/978-3-030-12767-1_5
[23] M. F. F. Ab. Rashid, "Tiki-taka algorithm: A novel metaheuristic inspired by football playing style", *Engineering Computations*, **vol. 38**, no. 1, 2021, pp. 313–343.
[24] S. Li, W. Gong, X. Yan, C. Hu, D. Bai, and L. Wang, "Parameter estimation of photovoltaic models with memetic adaptive differential evolution", *Solar Energy*, **vol. 190**, 2019, pp. 465–474.
[25] J. Wang et al., "Photovoltaic cell parameter estimation based on improved equilibrium optimizer algorithm", *Energy Conversion and Management*, **vol. 236**, 2021, p. 114051.
[26] A.-E. Ramadan, S. Kamel, T. Khurshaid, S.-R. Oh, and S.-B. Rhee, "Parameter extraction of three diode solar photovoltaic model using improved grey wolf optimizer", *Sustainability*, **vol. 13**, no. 12, 2021, p. 6963.
[27] H. Shaban et al., "Identification of parameters in photovoltaic models through a runge kutta optimizer", Mathematics, **vol. 9**, no. 18, 2021, p. 2313.
[28] M. Mostafa, H. Rezk, M. Aly, and E. M. Ahmed, "A new strategy based on slime mould algorithm to extract the optimal model parameters of solar PV panel", *Sustainable Energy Technologies and Assessments*, **vol. 42**, 2020, p. 100849.

[29] X. Ye et al., "Modified whale optimization algorithm for solar cell and PV module parameter identification", *Complexity*, 2021, pp. 1–23.
[30] M. H. Hassan, S. Kamel, M. El-Dabah, and H. Rezk, "A novel solution methodology based on a modified gradient-based optimizer for parameter estimation of photovoltaic models", *Electronics*, **vol. 10**, no. 4, 2021, p. 472.

Chapter 4

Comparative Analysis of Conventional and Cuckoo Search MPPT Algorithms in PV Systems under Uniform Condition

Abdelfettah El-Ghajghaj, Ouadiâ Chekira, Najib El Ouanjli, Hicham Karmouni, and Mhamed Sayyouri

4.1 INTRODUCTION

Currently, the global demand for energy is increasing exponentially due to the increase in population and industrial activities, which increases the pressure on natural energy resources by burning a lot of non-renewable fossil fuels. Therefore, the high use of fossil energy in production disturbs the climate balance due to the emission of greenhouse gases such as CO_2 [1, 2].

Renewable energies offer an indispensable solution to fight against pollution while meeting the great demand for energy, such as biomass, wind, and solar [3]. In this context, solar energy is among the most interesting sources because it has a better yield and availability [4, 5]. Photovoltaic panels (PVs) convert solar radiation into electricity, and then the power supplied by the panels is nonlinear. In addition, changing climatic conditions affect the energy efficiency of electric power generation. Indeed, the Power-Voltage (P-V) curve has a single maximum power point (MPP), which must be followed continuously by control techniques based on MPP tracker (MPPT) [6]. Several MPPT techniques exist in the literature, such as the classical methods including the perturb and observe (P&O), incremental conductance (IC), and hill climbing (HC) [7–11]. These classical methods find the MPP with better performance in terms of simplicity and speed, but the power shows oscillations around the MPP. Therefore, researchers propose several works based on metaheuristic methods to overcome the problem related to oscillations, such as particle swarm optimization (PSO) [12], grey wolf optimization (GWO) [13], and cuckoo search optimization [14]; these techniques minimize the oscillations, and at the same time the complexity remains average compared to other techniques based on genetic algorithm (GA) [15].

Among these metaheuristic methods, cuckoo search optimization provides a reliable solution for energy production by photovoltaic systems [16]. The CS method is able to detect local maximum power points induced by partial shading, thanks to its global search process and its ability to avoid local optimums. Additionally, CS can adapt to various environmental conditions, making it a robust method for monitoring MPP in dynamic environments.

DOI: 10.1201/9781003462460-4

Unlike some MPPT methods that require complex system modeling, CS is relatively simple to implement and does not require detailed knowledge of the photovoltaic system.

In this work, the classical P&O method and the metaheuristic CS method are presented and compared through a theoretical study and a simulation test using MATLAB®/Simulink®. The analysis of the results obtained clearly shows the efficiency of the CS algorithm compared to the P&O in terms of the oscillation problem and MPP tracking, which improves energy production.

The rest of the chapter is structured as follows: Section 4.2 focuses on the presentation of the system studied which is composed of a PV generator and a boost converter. The P&O and CS MPPT techniques are presented in detail in Section 4.3. The results of this study are presented and discussed in Section 4.4. A conclusion is presented in Section 4.5.

4.2 PV SYSTEM DESCRIPTION

The photovoltaic conversion chain is given in Figure 4.1, which is composed of four PV panels (Advance Power API P-215) and a boost converter connected to a load [17]. Under standard conditions, each PV panel generates up to 215 W.

The measurement of the voltage and current at the output of the PV panel allows the MPP to be monitored by the MPPT controller. In addition, the converter can be connected to continuous loads such as batteries, loads, or resistors.

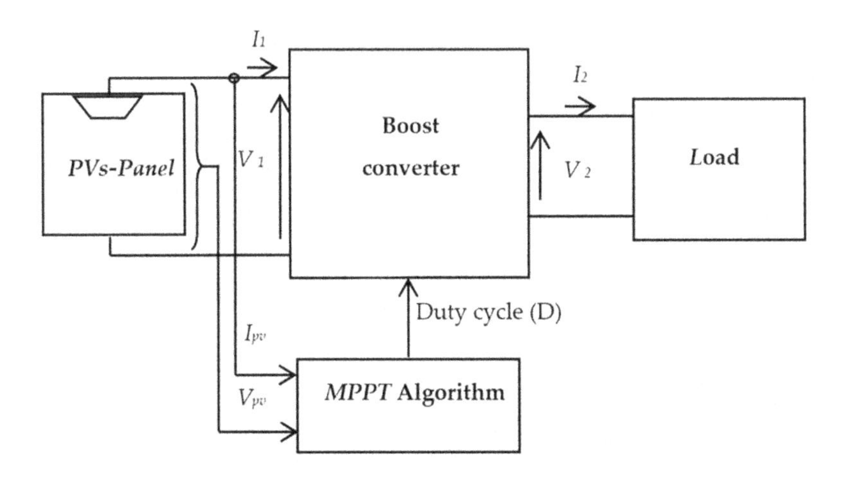

Figure 4.1 PV system installation.

Each photovoltaic cell of the PV panel can be represented by an equivalent circuit il-lustrated in Figure 4.2, where I_{ph-c} is the photocurrent, I_{pv-c} and V_{pv-c} are respectively the output current and voltage of PV cell, and the series and parallel resistances are illustrated by R_s and R_p.

The current I_{pv-c} and I_{ph-c} can be expressed by the following two relations [18, 19]:

$$I_{pv\_c} = I_{ph} - I_0\left[\exp\left(\frac{V_{pv\_c} + R_s I}{aKTN_s}\right) - 1\right] - \left(\frac{V_{pv\_c} + R_s I_{pv\_c}}{R_p}\right) \tag{4.1}$$

$$I_{ph\_c} = \left[I_{SC} + \alpha\,(T - T_r)\right]\frac{G}{1000} \tag{4.2}$$

The boost converter presented in Figure 4.3 plays a crucial role in linking the PV modules and the load. Its main purpose is to ensure consistent operation at peak power point by employing switching pulses generated through PWM. The essential components in the design encompass the inductance and capacitance integrated within the MPPT converter.

The relationship between input voltage (Vs) and output voltage (V_0) of the power interface is given as follows:

$$V_0 = \frac{V_s}{1 - D} \tag{4.3}$$

The relationship between input current (I_s) and output current (I_0) of the power interface is given as:

$$I_0 = I_s.(1 - D) \tag{4.4}$$

where: D is the duty cycle.

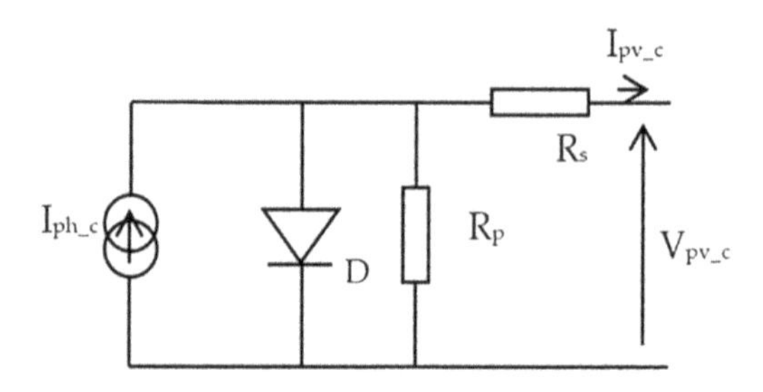

Figure 4.2 PV cell.

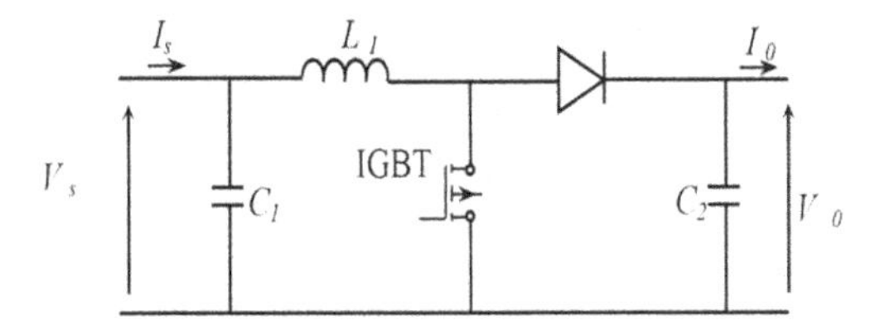

Figure 4.3 DC-DC boost converter.

4.3 MPPT TECHNIQUES

The choice of the appropriate control technique in conjunction with the MPPT algorithm plays a vital role in tracking the maximum power of the photovoltaic system. In this study, the direct control method is utilized, where the control signal is the duty cycle directly applied to the switching component of the power interface.

This section presents the MPPT techniques used in this chapter, such as P&O and CS-based MPPT; the theoretical basis of each technique is presented in detail.

4.3.1 P&O-Based MPPT

The perturb and observe (P&O) algorithm is notable for its inherent simplicity, cost-efficient implementation, and ease of execution. Moreover, a distinct advantage is its independence from the need for pre-existing knowledge about the characteristics of photovoltaic systems. This technique consists in provoking a small value perturbation on the voltage V_{pv} by generating a duty cycle, which results in a power variation. Figure 4.4 illustrates the detailed mechanism of the maximum power search process by the P&O algorithm [20].

4.3.2 CS-Based MPPT

The cuckoo search optimization method is an optimization technique inspired by the laying and incubation behavior of cuckoo eggs [21]. To tune the CS for MPPT, several appropriate parameters must be chosen for the search, such as duty cycle values, step size identified by α and fitness (J) which is the power at MPP to track the MPP a new duty cycle is created by the following equation:

$$D_i^{t+1} = D_i^t \oplus Levy(\lambda) \tag{4.5}$$

Where: $\alpha = \alpha_0\,(D_{best} + D_i)$

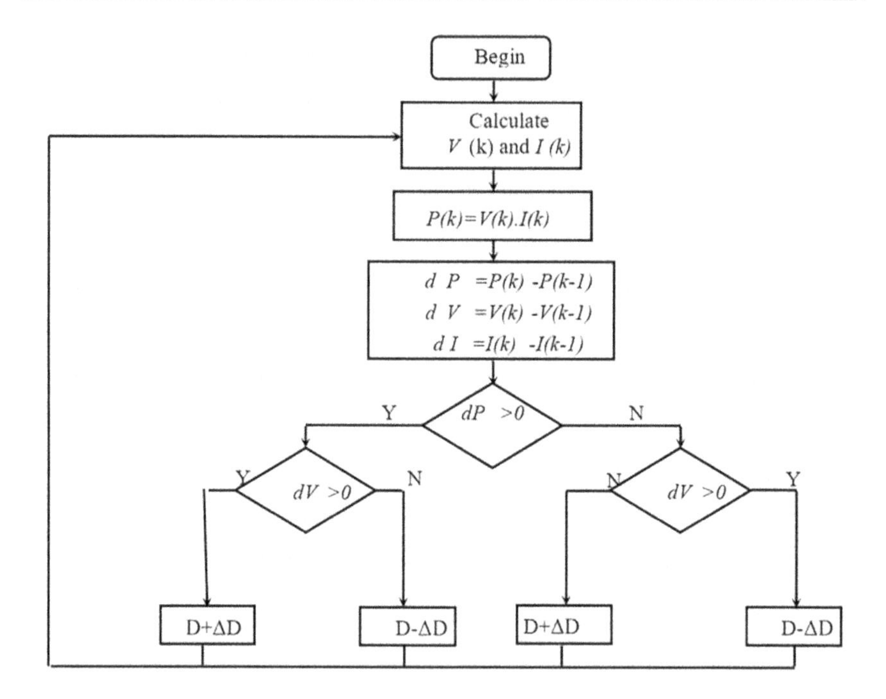

Figure 4.4 P&O-based MPPT.

The Levy allocation is presented as:

$$S = \alpha_0 (D_{best} + D_i) \oplus levy(\lambda) \approx k \times \left(\frac{u}{|v|^{\frac{1}{\beta}}} \right) (D_{best} + D_i) \tag{4.6}$$

Where: $\beta=1.5$, k is the Levy factor, where u and v are resolute from the distribution curves.

$$u \approx N(0, \sigma_u^2) \;\; v \approx N(0, \sigma_v^2) \tag{4.7}$$

The variable σ_u and σ_v are given by:

$$\sigma_u = \left(\frac{\Gamma(1+\beta) \times \sin\left(\pi \times \frac{\beta}{2}\right)}{\Gamma\left(\frac{1+\beta}{2}\right) \times \beta \times (2)^{\left(\frac{\beta-1}{2}\right)}} \right)^{\frac{1}{\beta}}, \quad \sigma_v = 1 \tag{4.8}$$

The detailed steps to search the MPP by CS-based MPPT are presented in Figure 4.5.

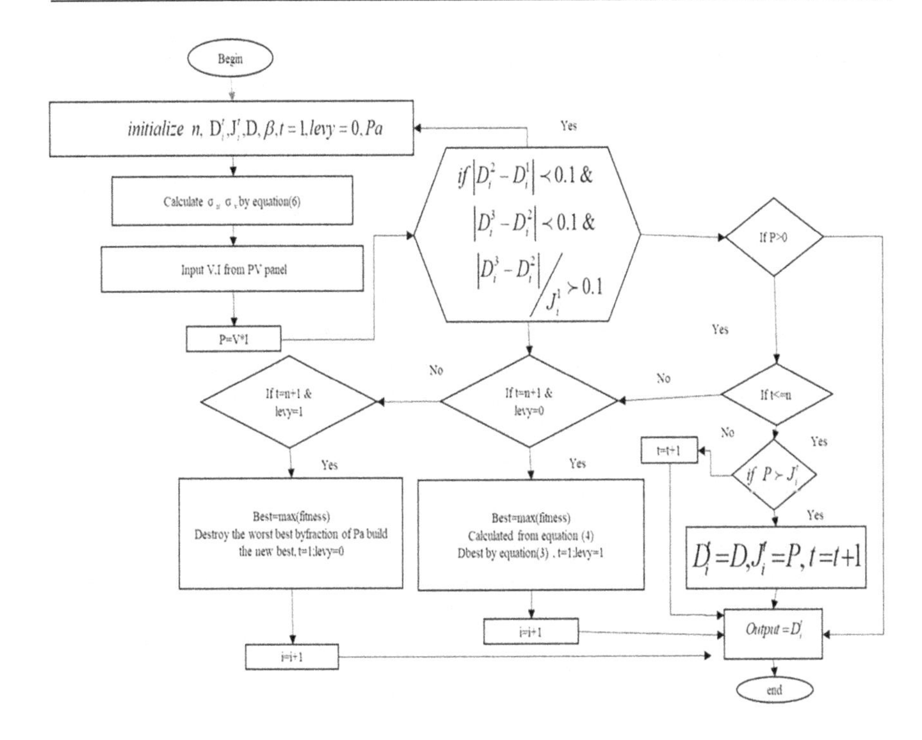

Figure 4.5 CS-based MPPT.

4.4 SIMULATION RESULTS AND DISCUSSION

To evaluate the effectiveness of the optimization techniques studied, a simulation test on MATLAB/Simulink software is applied on the PV system. This system is composed of four PV panels (Advance Power API P-215) connected in parallel and a boost converter linked to a resistor. The specifications of the PV panel employed are outlined in Section 4.6.

The test is carried out under a constant temperature of 25 degrees, and under variable irradiation as shown in Figure 4.6. During the first-time interval, the solar irradiance is 400 W/m^2. It increases to 500 W/m^2 during the second time interval, and further increases to 600 W/m^2 during the third time interval.

Figure 4.7 depicts the performance of a PV module under varying solar irradiance and constant temperature conditions. These curves highlight the significant influence of solar irradiance fluctuations on the power output, as demonstrated by the I-V and P-V curves of the four photovoltaic panels employed in this study. The P-V curve clearly shows the maximum power for each irradiation that must be followed to improve the production efficiency.

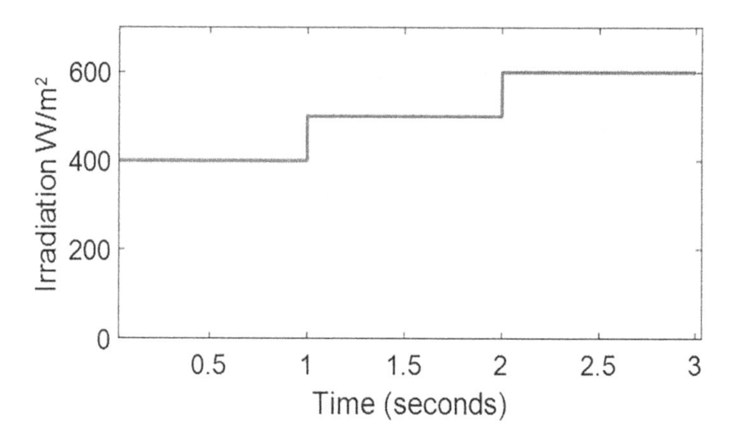

Figure 4.6 Irradiation condition.

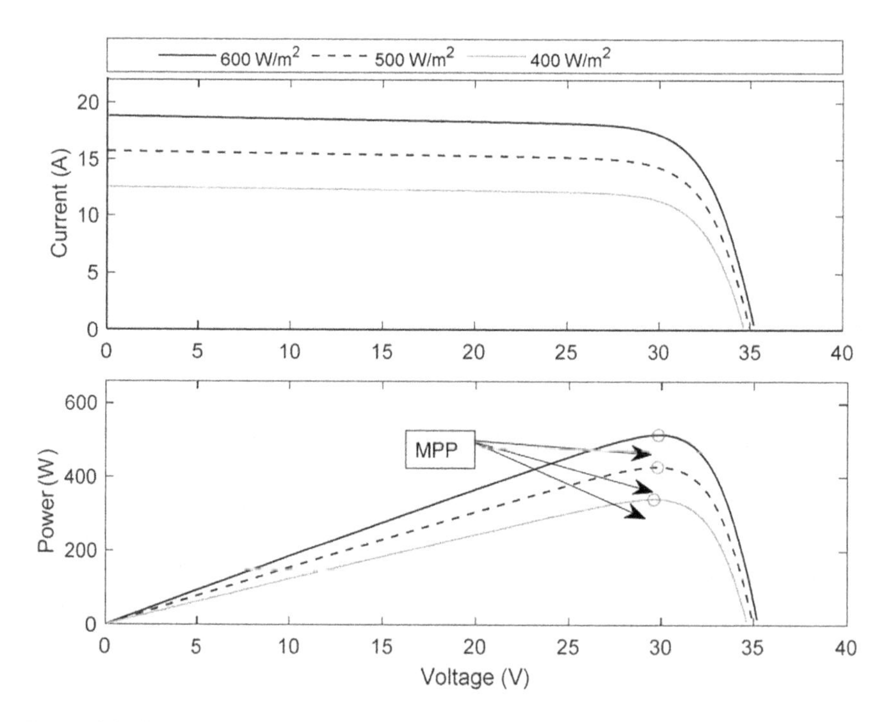

Figure 4.7 (I-V) and (P-V) curves.

Figures 4.8, 4.9 and 4.10 show the simulation results obtained. The power tracking by the P&O technique is shown in Figure 4.8; this method reaches up to 95% of the maximum power. However, the power found contains oscillations, moreover at each irradiation change, the MPP tracking is done by a small delay, which means a loss of energy.

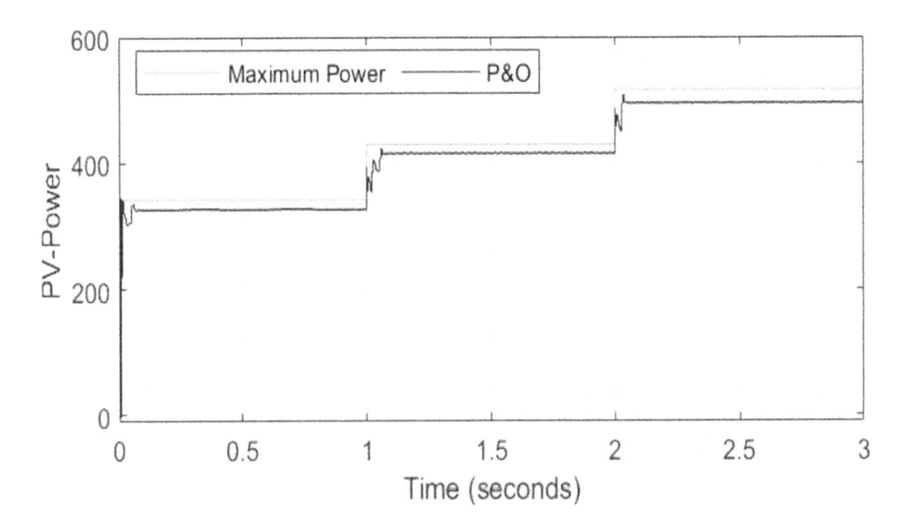

Figure 4.8 Tracking power using P&O.

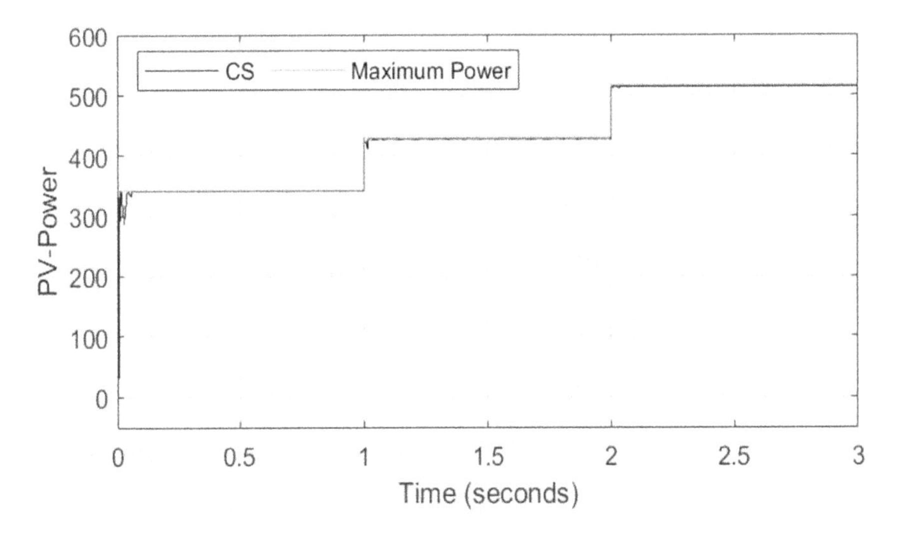

Figure 4.9 Tracking power using CS.

On the other hand, the CS-based MPPT method offers a better result on minimizing oscillations around MPP tracking reaching up to 99% of maximum power as shown in Figure 4.9.

In another way, the CS algorithm achieves maximum power tracking with minimal oscillation compared to the P&O technique, and it demonstrates no delay when irradiation levels change, as illustrated in Figure 4.10.

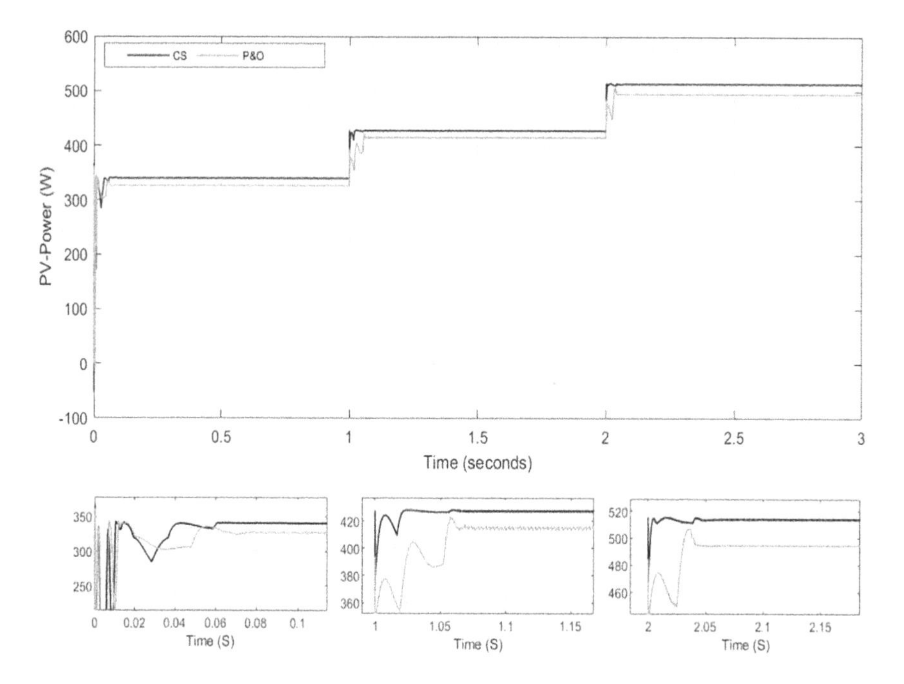

Figure 4.10 Tracking power using P&O and CS-based MPPT.

Table 4.1 summarizes the performances of the two optimization strategies. It is well observed that the statistics presented in this table show that the metaheuristic technique cuckoo search presents a better performance and efficiency compared to the classical technique P&O.

As can be observed, at the first, second, and third irradiations, the maximum PV power is approximately 342 W, 429 W, and 516 W respectively. The CS algorithm reached the maximum power of the PV array, achieving 340 W, 426 W, and 513 W respectively. However, the traditional P&O method approached a point close to the maximum power, reaching 325 W, 413 W, and 495 W respectively. Furthermore, the CS method demonstrates a higher speed in reaching maximum power compared to the P&O method.

4.5 CONCLUSION

This chapter gave a comparative analysis involving two MPPT algorithms – P&O and cuckoo search. The performance of each technique was tested using a photovoltaic system operating at constant temperature and variable solar irradiation. The simulation results in this work clearly show the effectiveness of the metaheuristic CS technique compared to the conventional

Table 4.1 Comparison Results between P&O and SC Technique

Irradiation (W/m²)		400	500	600
Power (W)	MPP	342	429	516
	P&O	325	413	495
	CS	340	426.5	513
Tracking efficiency %	P&O	95	96.2	95.9
	CS	99.4	99.3	99.4
Tracking speed (S)	P&O	0.08	0.07	0.05
	CS	0.06	0.026	0.04
Oscillations	P&O	Medium	Medium	Medium
	CS	Faible	Faible	Faible

P&O technique in terms of tracking and extracting power from the PV array and steady-state power oscillations.

4.6 APPENDIX

Table 4.2 Characteristics of the Advanced Power API-P215 PV Panel

Parameter	Value
Maximum Power (P_m)	214.96 W
Open circuit voltage (V_{oc})	36 V
Voltage at P_m (V_{mp})	29.94 V
Temperature coefficient of (V_{oc})	-(0.3342) %/deg.C
Cells per module	60
Short-circuit current (I_{sc})	7.83 A
Current at P_m (I_{mp})	7.18 W
Temperature coefficient of (I_{sc})	(0.057586) %/deg.C
Light-generated current (I_L)	7.85 A
Diode saturation current (I_0)	1.0301e-10 A
Diode ideality factor	0.93379
Shunt resistance (R_{sh})	98.7209 ohms
Series resistance (R_s)	0.23044 ohms

REFERENCES

[1] Jaramillo, P., & Muller, N. Z. (2016). Air pollution emissions and damages from energy production in the US: 2002–2011. *Energy Policy*, 90, 202–211.

[2] El Mourabit, Y., Derouich, A., Allouhi, A., El Ghzizal, A., El Ouanjli, N., & Zamzoumyes, O. (2020). Sustainable production of wind energy in the main Morocco's sites using permanent magnet synchronous generators. *International Transactions on Electrical Energy Systems*, 30(6), e12390.

[3] Mohtasham, J. (2015). Renewable energies. *Energy Procedia*, 74, 1289–1297.

[4] Capellán-Pérez, I., De Castro, C., & Arto, I. (2017). Assessing vulnerabilities and limits in the transition to renewable energies: Land requirements under 100% solar energy scenarios. *Renewable and Sustainable Energy Reviews*, 77, 760–782.
[5] Errouha, M., Derouich, A., El Ouanjli, N., & Motahhir, S. (2020). High-performance standalone photovoltaic water pumping system using induction motor. *International Journal of Photoenergy*, 1–13.
[6] Podder, A. K., Roy, N. K., & Pota, H. R. (2019). MPPT methods for solar PV systems: A critical review based on tracking nature. *IET Renewable Power Generation*, 13(10), 1615–1632.
[7] Femia, N., Petrone, G., Spagnuolo, G., & Vitelli, M. (2004, June). Optimizing sampling rate of P&O MPPT technique. In *2004 IEEE 35th Annual Power Electronics Specialists Conference* (IEEE Cat. No. 04CH37551) (Vol. 3, pp. 1945–1949). IEEE.
[8] El-Ghajghaj, A., Ouanjli, N. E., Karmouni, H., Jamil, M. O., Qjidaa, H., & Sayyouri, M. (2022). An improved MPPT based on maximum area method for PV system operating under fast varying of solar irradiation. In *Digital Technologies and Applications: Proceedings of ICDTA'22, Fez, Morocco* (Vol. 1, pp. 545–553). Springer International Publishing.
[9] Hilali, A., El Ouanjli, N., Mahfoud, S., Al-Sumaiti, A. S., & Mossa, M. A. (2022). Optimization of a solar water pumping system in varying weather conditions by a new hybrid method based on fuzzy logic and incremental conductance. *Energies*, 15(22), 8518.
[10] Safari, A., & Mekhilef, S. (2011, May). Incremental conductance MPPT method for PV systems. In *2011 24th Canadian Conference on Electrical and Computer Engineering* (CCECE) (pp. 000345–000347). IEEE.
[11] Rawat, R., & Chandel, S. (2013). Hill climbing techniques for tracking maximum power point in solar photovoltaic systems-a review. *International Journal of Sustainable Development and Green Economics (IJSDGE)*, 2, 90–95.
[12] Ishaque, K., Salam, Z., Amjad, M., & Mekhilef, S. (2012). An improved particle swarm optimization (PSO)–based MPPT for PV with reduced steady-state oscillation. *IEEE Transactions on Power Electronics*, 27(8), 3627–3638.
[13] Mohanty, S., Subudhi, B., & Ray, P. K. (2015). A new MPPT design using grey wolf optimization technique for photovoltaic system under partial shading conditions. *IEEE Transactions on Sustainable Energy*, 7(1), 181–188.
[14] Ahmed, J., & Salam, Z. (2014). A Maximum Power Point Tracking (MPPT) for PV system using Cuckoo Search with partial shading capability. *Applied Energy*, 119, 118–130.
[15] Borni, A., Abdelkrim, T., Bouarroudj, N., Bouchakour, A., Zaghba, L., Lakhdari, A., & Zarour, L. (2017). Optimized MPPT controllers using GA for grid connected photovoltaic systems, comparative study. *Energy Procedia*, 119, 278–296.
[16] Hilali, A., Mardoude, Y., Essahlaoui, A., Rahali, A., & El Ouanjli, N. (2022). Migration to solar water pump system: Environmental and economic benefits and their optimization using genetic algorithm based MPPT. *Energy Reports*, 8, 10144–10153.
[17] Awasthi, A., Shukla, A. K., SR, M. M., Dondariya, C., Shukla, K. N., Porwal, D., & Richhariya, G. (2020). Review on sun tracking technology in solar PV system. *Energy Reports*, 6, 392–405.

[18] El-Ghajghaj, A., Karmouni, H., El Ouanjli, N., Jamil, M. O., Qjidaa, H., & Sayyouri, M. (2023, January). Comparative analysis of classical and meta-heuristic MPPT algorithms in PV systems under uniform condition. In *International Conference on Digital Technologies and Applications* (pp. 714–723). Springer Nature Switzerland.
[19] da Rocha, M. V., Sampaio, L. P., & da Silva, S. A. O. (2020). Comparative analysis of MPPT algorithms based on Bat algorithm for PV systems under partial shading condition. *Sustainable Energy Technologies and Assessments*, 40, 100761.
[20] Motahhir, S., El Ghzizal, A., Sebti, S., & Derouich, A. (2015, December). Proposal and implementation of a novel perturb and observe algorithm using embedded software. In *2015 3rd International Renewable and Sustainable Energy Conference (IRSEC)* (pp. 1–5). IEEE.
[21] Peng, B. R., Ho, K. C., & Liu, Y. H. (2017). A novel and fast MPPT method suitable for both fast changing and partially shaded conditions. *IEEE Transactions on Industrial Electronics*, 65(4), 3240–3251.

Chapter 5

Overview of Solar Concentrator Technologies and Their Role in the Development of CSP Plants

Firyal Latrache, Najwa Jbira, Zakia Hammouch, K. Lamnaouar, Benaissa Bellach, and Mohammed Ghammouri

5.1 INTRODUCTION ON SOLAR CONCENTRATORS AND CSP PLANTS

Since economic activity was transformed during the Industrial Revolution, energy consumption has continued to rise. Certainly, before the Industrial Revolution, people consumed "this energy" by combusting wood, gas and even coal to heat their homes and satisfy their daily needs. Indeed, the Industrial Revolution introduced the steam engine, which was considered the great innovation of this revolution, followed by other inventions: the telegraph (1791), gas lighting (1786), the steam locomotive (1829) and photography (1839). It is obvious that before, during and after the Industrial Revolution, the world's energy consumption was based entirely on the triple "wood, gas and coal", which had to be combined to generate thermal energy (for heating) and mechanical energy (to power internal and external combustion engines). The combustion of these fossil resources produces carbon dioxide and greenhouse gases, which in turn cause environmental degradation and global warming. To this end, the use of renewable energies such as solar energy, wind power, geothermal energy and hydropower reduce the impact of air pollution and therefore preserve the environment.[1–4]

In fact, global energy consumption has reached 605 exajoules by 2022, with carbon dioxide emissions from the energy sector at 87%. This reflects the widespread use of fossil fuels, despite strategies to increase the renewable energy contribution to the energy mix. With regard to the proportion of fossil and renewable resources in the global energy mix in 2022, oil, coal and natural gas widely contributed by 31.569%, 26.731% and 23.490% respectively. Renewable resources contribute only 6.734% for hydropower and 7.479% for other inexhaustible resources. This rising proportion of fossil resources shows that renewable energies do not make any diminution for the triple "gas, oil, coal" use. It is therefore recommended to undertake an energy transition towards renewable and efficient energy consumption to increase the proportion of renewable energies and protect the natural resources.[5–8]

DOI: 10.1201/9781003462460-5

As mentioned earlier, and in view of the existing renewable resources – wind energy, geothermal energy and hydropower – solar energy is used to generate electrical energy by converting solar radiation through systems. If these solar rays are converted into electrical energy using photovoltaic solar panels, the technology used is called "photovoltaic solar energy". If there are mirrors that concentrate the incident solar radiation and reflect it onto a receiver to heat the fluid in it to generate steam, the technology is called "solar thermal energy". If the steam produced drives a turbine, in this case the electricity is generated through "thermodynamic solar energy". This thermodynamic solar technology is part of the operating principle of CSP concentrating solar power plants: incident solar radiation is reflected by various geometries of solar concentrators to generate steam and then electrical energy, which is stored for later use. Then a CSP power plant is composed of a field of solar concentrators, a storage system and a conversion system. When solar radiation arrives at the solar concentrator, it is reflected towards the receiver, where a heat transfer fluid vaporizes and transports thermal energy to the conversion circuit. This thermal energy could be stored in a thermal storage system and later converted into electrical energy.[9–11]

Current solar concentrator geometries include parabolic solar concentrators, parabolic trough solar concentrators, heliostats, Fresnel mirrors, elliptical hyperboloid solar concentrators, Fresnel lens solar concentrators, compound parabolic solar concentrators, total internal reflection dielectric concentrators, quantum dot concentrators and conical solar concentrators. CSP plants are classified into three generations according to the solar concentrator geometry used and the efficiency of the thermodynamic cycle adopted to describe the process. In general, parabolic trough concentrators, heliostats and Fresnel mirrors are used in first- and second-generation CSP plants, with Rankine cycle efficiencies ranging from 28% to 38% for the first and from 38% to 44% for the second. Third-generation CSP plants, on the other hand, use more advanced materials to raise the operating temperature of first- and second-generation plants and also to optimize thermal storage using the Brayton cycle. Concerning the heat transfer fluids used in these CSP plants: first- and second-generation CSPs work with steam, oil and salt, while third-generation CSPs, developed to increase the efficiency of first- and second-generation plants, handle salt and a few gases such as helium and air. [12–14]

According to the International Energy Agency (IAE), CSP will cover 11.3% of global electricity demand by 2050, with full optimization of thermal storage. At present, Spain, the United States and China are the leaders in the installation of CSP plants, with production worth 2,300 MW, 1,500 MW and 906 MW respectively, followed by the UAE, South Africa and Morocco, where the CSP plants installed in Morocco (Noor I-IV Drâa-Tafilalet, Airlight Energy Ait Baha and Ain Beni Mathar) produce 533 MW of electricity.[13] These CSP plants have an advantage over photovoltaic solar energy and other renewable technologies in terms of thermal storage and other environmental benefits: limited access to water in the condensation phase compared with other fossil fuel power plants, which consume large quantities of water in the cooling towers. Furthermore, CSP plants do not cause climate change and do not require a lot of space for installation. Despite these beneficial effects, CSP can have harmful effects on the environment, biodiversity and human health, as well as noise pollution with some solar concentrator geometries.[15, 16]

To ensure that CSP plants contribute to electricity generation by 2050, third-generation CSP plants need to provide thermal storage for up to 8 hours to cover electricity demand.[13] Consequently, this generation of CSP must use thermal storage with a lower installation cost to guarantee economical and efficient use of CSP plants. In addition, third-generation CSPs must respect environmental biodiversity and amortize the noise arising from the use of certain solar concentrators.[17]

It is obvious that CSP plants require more demonstration to guarantee a high efficiency of electricity production with a thermal storage capacity of up to 8 hours.[18, 19] The aim of this chapter is therefore to help overcome this challenge so that CSPs can satisfy electricity demand in the future. This chapter presents and describes the geometries of the solar concentrators (previously listed) currently on the commercial market, and their broad applications. Indeed, this chapter will help researchers in this field to fully understand the geometries and discover their uses, and also to invest in third-generation CSP plants to increase the rate of electricity production and guarantee subsequent use of this electricity through lower-cost thermal storage. Then, it encourages them to build CSP power plants with other geometries that are not widely used: elliptical hyperboloid solar concentrators, Fresnel lens solar concentrators, compound parabolic solar concentrators, total internal reflection dielectric concentrators, quantum dot concentrators and conical solar concentrators.

The chapter starts by giving the criteria for the classification of solar concentrators, and then goes on to describe each solar concentrator geometry, showing its applications. At the end, the chapter analyses and discusses solar concentrators efficiency to enrich the literature and inspire investors and researchers to build new CSP plants with more efficient solar concentrators.

5.2 CLASSIFICATION OF SOLAR CONCENTRATORS

There are a several types of solar concentrator technologies that continue to capture the interest of researchers. This solar concentrator can be classified according to the following criteria:

- Optical characteristics: Concentrators can be categorized into imaging and non-imaging systems. Imaging concentrators can further be classified as linear or point concentrators, depending on their optical configuration.[20]

- Concentration ratio (C): The concentration ratio determines the operating temperatures of concentrator systems. Concentrators can be categorized as:[21]
  - Low concentrations (1 < C <10) with operating temperatures around 150°C.
  - Medium concentrations (C > 100) with operating temperatures around 300°C.
  - High concentrations (C > 100) with operating temperatures exceeding 500°C.
- Geometric characteristics: Concentrators can be distinguished based on their geometric configuration, including two-dimensional (2D) and three-dimensional (3D) systems. The geometric characteristics play a crucial role in determining the concentration and focusing of solar energy.[22]
- Pointing modes: Concentrators can have different pointing modes, which determine their orientation and tracking capabilities. Pointing modes can include fixed or periodically oriented systems, as well as systems that are mobile around one axis or mobile around two axes. These pointing modes affect the efficiency and adaptability of concentrator systems to track the sun's movement.[23]
- Relative positions of the absorber and concentrator: Concentrators can have varying arrangements in terms of the relative positions of the absorber and the concentrator itself. This can include systems where one component is movable relative to the other, or systems where the absorber and concentrator are integral and fixed in position. The arrangement impacts the efficiency and complexity of the concentrator system.[24]

5.3 SOLAR CONCENTRATOR TECHNOLOGIES

There are varieties of solar concentrator technologies that have been used extensively: parabolic concentrators, parabolic trough concentrators, Fresnel mirrors and heliostats. Other solar concentrator systems have also been developed, such as hyperboloid concentrators, Fresnel lens concentrators, compound parabolic concentrators, total internal reflection dielectric concentrators, quantum dot concentrators and conical concentrators. Each technology and geometry has its own operating principle, advantages and disadvantages, making them a subject for today's solar concentrator market.

The aforementioned technologies are divided into four categories: reflectors, refractors, hybrid reflector-refractor systems and luminescent. Parabolic concentrators, parabolic trough concentrators, compound parabolic concentrators, hyperboloid concentrators and conical concentrators are reflectors, while Fresnel lens concentrators are refractors. In addition, dielectric concentrators with total internal reflection constitute a hybrid reflector-refractor system. For quantum dot concentrators, photons undergo total internal reflection, and this type of concentrator is said to be luminescent, i.e. an object that emits light without producing heat. Luminescence is the phenomenon of light emission by the interaction between electrically charged particles.

5.3.1 Parabolic Solar Concentrators

Parabolic solar concentrators are designed as concave disc mirrors which reflect the solar radiation on a point called "focal point" and gives an output temperature between 600°C and 1000°C. At the focal point, there is the Stirling engine, which receives concentrated solar radiation. The Stirling engine operates due to the high temperatures and pressures of the heat transfer fluid contained in it and converts thermal energy into mechanical energy, which is then converted into alternating current by a generator (Figure 5.1). In fact, gas in Stirling engine is bounded between a hot piston (which is near to the lowest temperature value in the heat exchanger) and a cold piston (in front of the heat exchanger with a high temperature value). There are three types of Stirling engine: alpha, beta and gamma, and Table 5.1 presents the different technical characteristic of each type of Stirling engine. The alpha configuration is the most recent development of the Stirling solar engine, with

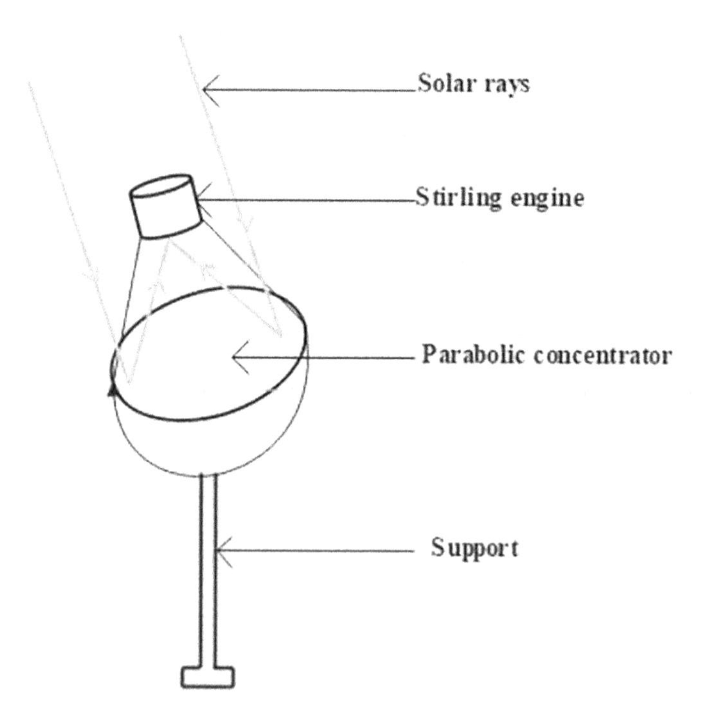

Figure 5.1 Parabolic solar concentrator with the Stirling engine.

Table 5.1 The Different Configurations of the Recently Developed Stirling Solar Engine [27, 29, 30]

Stirling Solar Motor Configuration	*Nominal Power (kW)*	*Work Pressure (MPa)*	*Maximum Gas Temperature (K)*	*Maximum Efficiency (%)*	*Speed (tr/min)*	*Work Fluid*
Alpha	4.1	0.2	877	39.5	3000	Hydrogen
Beta	0.1	2.85	771	12.1	800	Helium
Alpha	10	0.5–8	1243	46.67	1500	Helium
Alpha	3.65	5.3	849	26.2	1500	N/A
Alpha	26.6	10	923	26.9	1800	Helium
Alpha & Beta & Gamma	3	6.1	975	32	500–1500	Helium

a maximum gas temperature and speed of up to 1243 K and 1500 tr/min. Otherwise, the compact design of Beta type lead this configuration to have no aerodynamic losses, but it does generate a conduction heat one.

In general, a parabolic solar concentrator driven by a Stirling engine is operated in electricity generation, off-grid electrification, hybridization and storage, cooking, water distillation and desalination, and water pumping and irrigation. Experimental results of the implementation of a parabolic solar concentrator with a Stirling engine in grid-connected with irradiation density of 725 W/m^2 indicate that the system can produce 1 kW of electricity with an efficiency of 17.6%. In addition, an analysis of climate impact shows that a parabolic solar concentrator accompanied by Stirling engine achieves efficiencies of 24% and 50% respectively in desert and humid climates. The parabolic solar concentrator is applied with Stirling engine to realize a stand-alone system for off-grid electrification and demonstrates that reduces electricity costs by around 70% with an enhancement of system efficiency with 3%. Moreover, it is used as a solar cooker, which led water to reach 90°C between 90 and 120 minutes with an efficiency of 22%. The heat generated can be stored by a phase change material saline material as a phase change material with melting point of around 220°C which cooks French fries for just 17 minutes during a charging cycle.[25–33]

5.3.2 Parabolic Trough Solar Concentrators

In contrast to the parabolic solar concentrator, a parabolic through solar one collects and reflects the direct solar radiation onto a receiver tube placed in the focal line (Figure 5.2). The receiver in this case is an absorber tube that absorbs solar energy and converts it into thermal energy to heat and vaporize a heat transfer fluid. This heat transfer fluid drives the turbine in a thermodynamic cycle to generate electrical energy. In fact, the absorber tube is a

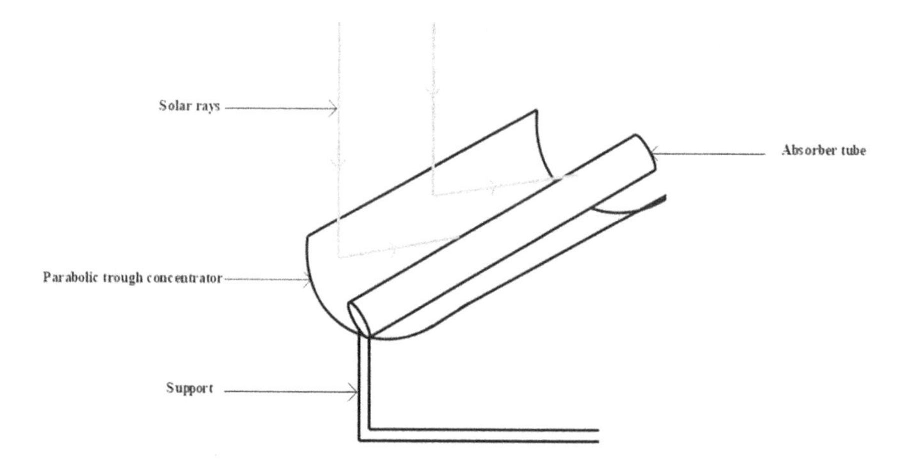

Figure 5.2 Parabolic trough solar concentrator with an absorber tube.

major element in the heat transfer process along the parabolic through solar concentrator, and it must have good absorptivity to limit the non-reflected solar radiation and therefore the heat losses. These heat losses are limited via selective surface coated on a metal tube. Among the existing types of selective surface: black metal oxide, multilayer anti-reflective coating, etc. In order to transport heat successfully, the heat transfer fluid must have low-pressure drop, high thermal capacity and conductivity, low viscosity, etc. Oils have a temperature range limited to 400°C, and molten salts based on sodium and potassium nitrates can reach 650°C. Furthermore, liquid water possesses an excellent heat transfer coefficient and can generate high temperatures, while organic fluids such as butane have a low evaporation temperature.

To improve the parabolic trough solar concentrator efficiency, recent research works focus on studying the impact of the optical errors and the heat flux distribution on the absorber tube. In this context, a microsystem for the production of thermal and electrical energies was tested. As a result, the electricity and thermal generation efficiency increased by 3% and 38% respectively. Even more, the parabolic through concentrator possesses an excellent operating performance in sunny and cloudy weather conditions with an efficiency of 70% in experimental tests. Therefore, the recent research work of parabolic through solar concentrator led applications in electricity generation and seawater desalination. There are other applications of parabolic trough in cooking water pasteurization processes. Other applications of the system such as solar heat generation in industrial and low-enthalpy processes and the production of chilled water can be added.[34–40]

5.3.3 Heliostats

The Greek words *Heliostats* is composed of *helios*, which means sun, and *stat*, which means stationary. Heliostats was developed in the 18th century by Fahrenheit. Later on, Foucault, Silbermann, Gambey and others improved it based on the clockwork mechanism and particularly Willem Jacob's Gravesande, which was developed so that can concentrate the incident solar radiation in a focal point. Similar to parabolic and parabolic trough solar concentrators, heliostats can generate electricity by focusing incident solar radiation onto a focal point located atop a tower where a receiver exists. Current applications use towers between 50 and 100 metres high, with receivers that are cavities or even doubles cavities led from an outer tube (Figure 5.3). Tower power plants may reach temperatures beyond 600°C by using molten salts and water. To follow the sun path, the heliostat has a tracking system, which is a mechanical component enabling the system to move. In fact, the altazimuth mount is the most widely used in the tracking system. It exists between two perpendicular axes: one vertical and the other horizontal. The heliostat rotates around the horizontal axis, which in turn carries the altazimuth mount to rotate around the vertical axis. For technological applications, heliostats are designed into two fields: heliostat fields that follow a geometric pattern and fields that do not. Geometric pattern heliostat fields are simple to construct, but the geometric rules impose certain limitations that lead to the establishment of forbidden zones where heliostats cannot be placed. However, non-geometrical pattern heliostats have no significant limitations, which allows them to reach excellent optical efficiency due to their adaptability. Among the heliostat field layouts based on a geometric pattern there are rectangular, staggered rectangular, radial, and staggered radial configurations. Regarding the non-geometric heliostat fields, they provide increased optical efficiency and better performance. They are, however, limited by a number of factors, including the tower's distance from the heliostats

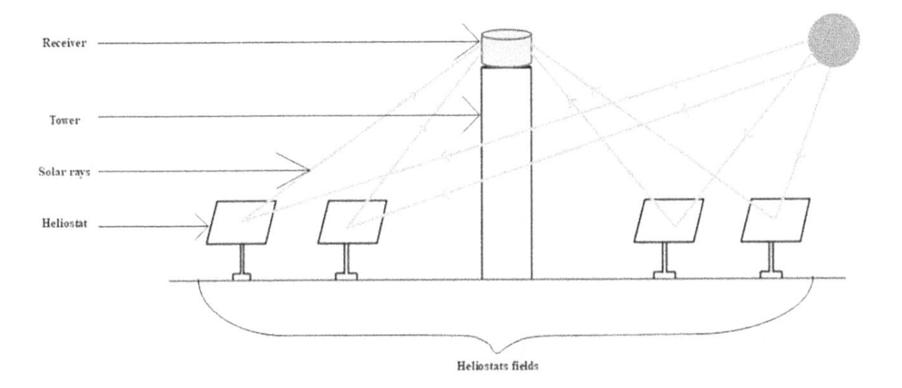

Figure 5.3 Heliostats fields with the solar tower.

and the heliostat field's size, which needs to be extended in order to reach the necessary minimum power. There are some algorithms such as JADE, SHADE and EB-LSHADE algorithm, have been used to optimize non-geometric heliostat fields to achieve the minimum power required.[41–47]

5.3.4 Fresnel Mirrors

Augustin Fresnel invented Fresnel mirrors in 1819 when studying the interference and polarization of light. Later in 20th century, this geometry inspired the Finland company Solar Fire Concentration to manufacture Lytefire solar concentration furnaces to reduce the use of fossil fuels in small-scale daily domestic activities. Then in 1963, Giovanni Francia developed the first Fresnel linear concentrator. Fresnel mirrors have a rectangular geometric form and collect the incident solar radiation to be concentrated on an absorber tube (Figure 5.4). This tube absorbs the solar energy and converts it into thermal energy to vaporize the existing heat transfer fluid and then to generate electricity. Fresnel mirrors offer numerous advantages: they can be installed on the ground or on sloping roofs, are easy to maintain and clean, and can be easily integrated into heating systems. The literature presents several Fresnel mirror configurations that contribute to increase the energy efficiency of the system: conventional Fresnel mirror layout and compact Fresnel mirror designs. The conventional Fresnel mirror possesses optimal angles of inclination to concentrate the maximum quantity of solar radiation onto an absorber tube. While compact configurations are designed to avoid the effects of blocking and shading. As a result, the conventional scheme will deliver the best performance, compared with the two other designs, which result in high light-beam scattering due to the location of the reflectors away from the

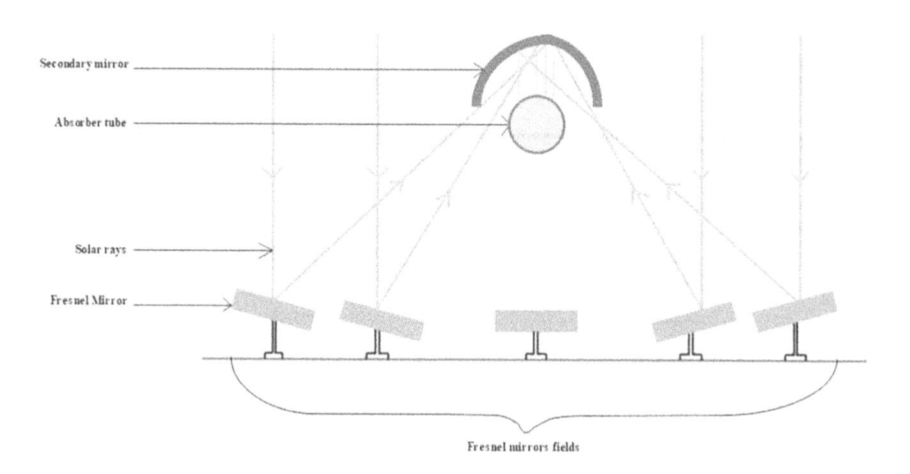

Figure 5.4 Fresnel mirrors field with an absorber tube and a secondary mirror.

absorber tube. As the reflectors are positioned away from the absorber tube, the conventional scheme will operate better than the other two designs, which lead to more light-beam scattering. To reduce the portion of fossil resources in the energy mix, Fresnel mirrors are applied in power generation, cogeneration, refrigeration, etc. The Fresnel mirror is applied also as a main heat source, a cogeneration system and as source to power a double-acting absorption chiller. Applied as a support system instead of using it as a primary heat source, Fresnel mirrors system generate electricity with an efficiency of 25%. Combined with a gas turbine, the Fresnel mirror system generates electricity with an efficiency of around 85%. Using of Fresnel mirror as a power source to a double-acting absorption chiller led to a performance coefficient COP of between 1.10 and 1.25 with an efficiency of up to 75%.[49–53]

5.3.5 Elliptical Hyperboloid Concentrator

The elliptical hyperboloid concentrator was first introduced in 2008 with an analysis of its elliptical plane, which moves with the sun path in the north-south direction each year and in the east-west direction each day. The elliptical hyperboloid concentrator has a large aperture angle and consequently high optical efficiency. It collects all incident solar radiation density to be concentrated on the outer surface of the receiver (Figure 5.5). The majority of elliptical hyperboloid applications focus on water desalination, which is used with helical receiver for water desalination. This application lead to reach an efficiency of about 50% with a concentration ratio varying between 7 and 8. Furthermore, the efficiency of the system is evaluated by using TiO_2 as a nanofluid to improve heat transfer in the receiver.[55–58]

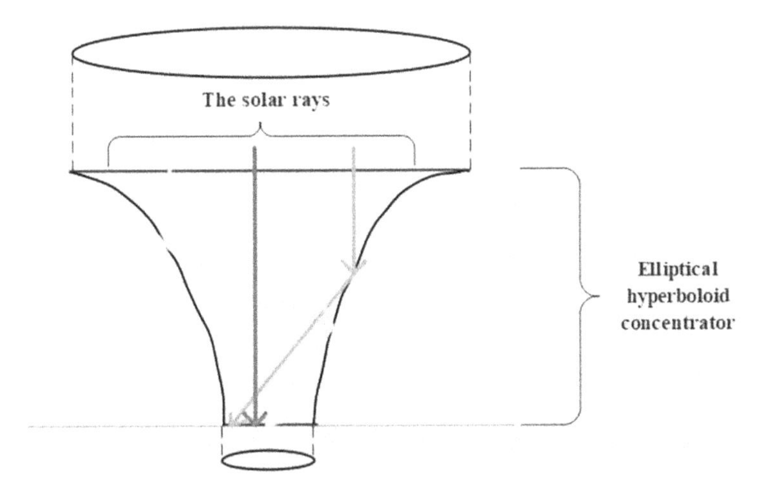

Figure 5.5 Elliptical hyperboloid solar concentrator.

5.3.6 Fresnel Lens Solar Concentrator

Fresnel lens solar concentrator technology first appeared in 1990 and it is composed of two parts: a flat top surface and a back surface with inclined facets. When solar radiation arrives at the top surface, it will be refracted and then concentrated at a focal point; this is the operating principle of conventional lens (Figure 5.6). The advantage of the Fresnel lens is that it is thinner and requires less material to manufacture. However, it suffers from some errors in the manufacturing process, which can result in a rounded geometrical form and then the receiver will not absorb all the incident solar radiation. Actually the research works are focused on the introduction of the non-imaging elliptical-based Fresnel lens concentrator. The Fresnel lens solar concentrator is applied in concentration of solar radiation, condensation, photoelectric, photovoltaics, solar control in buildings, etc. In this context, photovoltaic modules placed in the focal zone of stationary Fresnel mirror concentrators have been installed to satisfy the energetic needs in the buildings. As a result, this integrated system reaches a high percentage of 50% compared with conventional devices (simply fixed photovoltaic modules). The currently developed Fresnel lens solar concentrators are the Fresnel lens solar concentrator with test, the Fresnel lens solar laser and the Fresnel lens dome.[59–63]

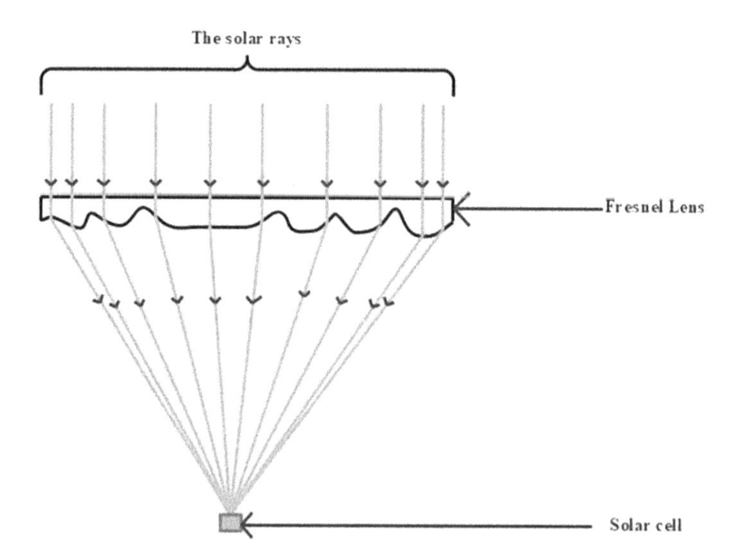

Figure 5.6 The use of the Fresnel lens to concentrate the incident solar radiation on a solar cell.

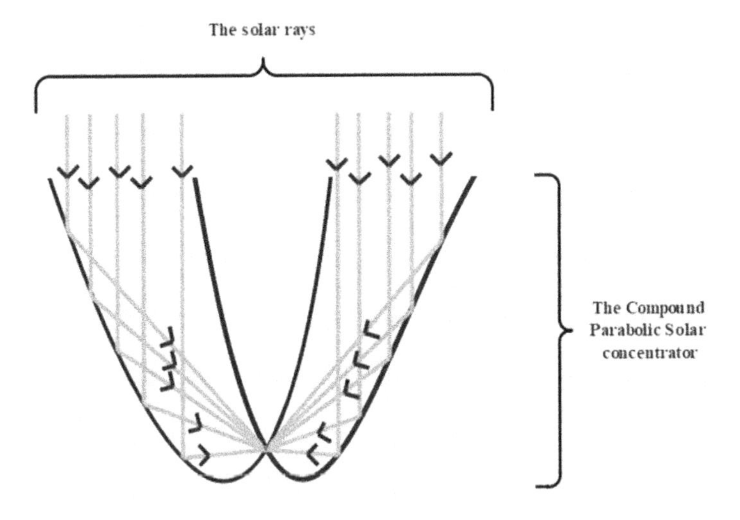

Figure 5.7 Compound parabolic solar concentrator.

5.3.7 Compound Parabolic Solar Concentrator

The compound parabolic solar concentrator was developed by S.p.A Solar Heat and power and used as a commercial product in solar energy applications. In general, a compound parabolic concentrator is classified into two types: two-dimensional (2D) and three-dimensional (3D). A 2D compound parabolic concentrator is a linear system with a longitudinal axis (Figure 5.7). Whereas a 3D system is obtained by rotating a meridian of the 2D compound parabolic concentrator at an angular interval. The optical performance of 3D compound parabolic solar concentrator depends on the concentrator apertures sides: more sides get ideal transmission and high optical efficiency. Because of its high concentration rate, the application of compound parabolic solar concentrator is widely used in photovoltaic/thermal hybrid systems and as a heat source in water heater system with storage. The collection efficiency in this last system achieves 53%. Energy production is maximized in a small-scale plant, where the compound parabolic solar concentrator operates an organic Rankine cycle and is used as a heat source, when the concentration ratio varies between 1.1 and 1.4 and the angle of inclination is less than 20°–25°. Generally the compound solar parabolic concentrator is compound with Fresnel mirrors to be applied in steam generation and water distillation.[64–70]

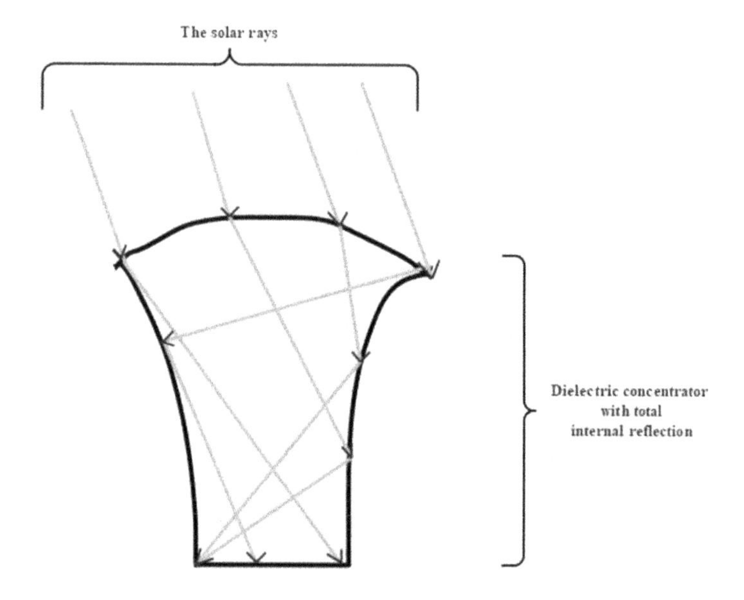

Figure 5.8 Dielectric concentrator with total internal reflection.

5.3.8 Dielectric Concentrator with Total Internal Reflection

The first design of dielectric concentrator with total internal reflection was introduced in 1987, and it is currently used as a secondary concentrator for space solar thermal applications at NASA. A total internal reflection dielectric concentrator is one of the technologies currently available that enables the achievement of concentrations ratios near to the theoretical maximum limits (Figure 5.8). In addition, this concentrator system provides a high concentration ratio with smaller aperture angle.[71–74]

5.3.9 Quantum Dot Concentrator

Quantum dot, the incredible particles discovered by the three scientists – Moungi Bawendi, Louis Brus and Alexei Ekimov – who won in 2023 the Nobel prize in chemistry. In fact, the quantum dot concentrator is composed of three parts: a transparent sheet of glass or plastic doped with quantum dots, reflective mirrors and a photovoltaic cell attached to the output (Figure 5.9). When incident solar radiation strikes the concentrator surface, it is refracted by the fluorescent material and absorbed by the quantum dots. The photons are then directed to the photovoltaic cell by total internal reflection. Moreover, the development of a quantum dot concentrator is limited by certain constraints related to luminescent colorants.[75]

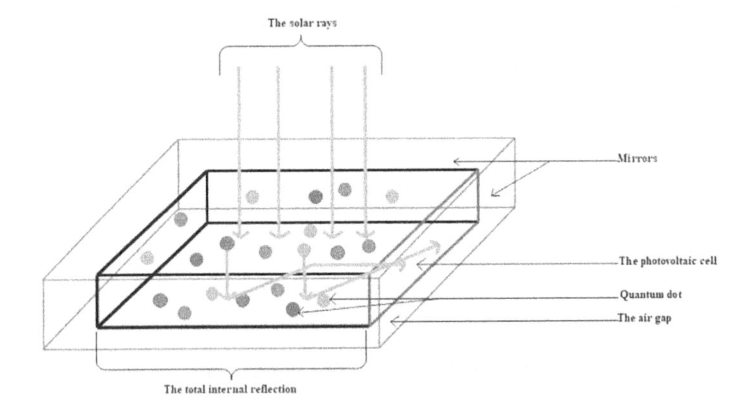

Figure 5.9 Quantum dot solar concentrator.

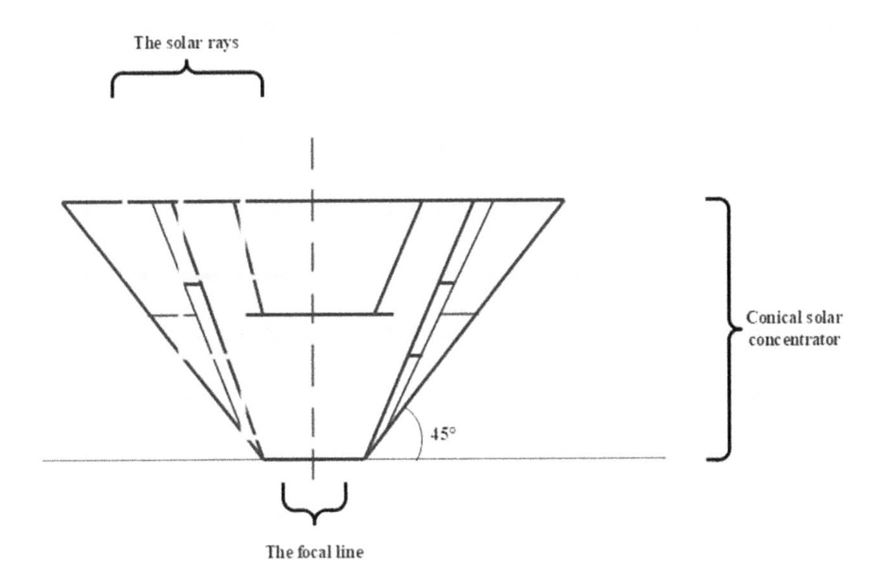

Figure 5.10 Conical solar concentrator with the conical angle of 45°.

5.3.10 Conical Solar Concentrator

The conical solar concentrator is one of the geometries, which collects and concentrates incident solar radiation on a receiver, placed at the focal line, to convert solar energy into electricity to satisfy energy needs in domestic and industrial applications. According to current research works, a conical angle of 45° maximizes the concentrated sun rays and therefore minimizes heat losses (Figure 5.10). Conical solar concentrators still need research contributions despite the ample literature aiming to develop solar concentrator technologies, as most recent studies have focused on parabolic and parabolic trough geometries, heliostats and Fresnel mirrors. The distribution of direct solar radiation, ambient temperature, inlet temperature, mass flow rate of the heat transfer fluid and solar concentrator concentration ratio can all be looked at and improved to evaluate the efficiency of a conical solar concentrator.[76]

5.4 ANALYSIS AND COMPARISON OF SOLAR CONCENTRATORS TECHNOLOGIES

The following tables, Table 5.2, Table 5.3, Table 5.4, Table 5.5 and Table 5.6, compare between the afore-presented solar concentrators geometries, either integrated in CSP plant or not. These comparisons are based on the efficiencies, applications and the operating temperatures of the solar concentrators technologies. For parabolic, trough parabolic, heliostats and Fresnel mirrors, the tables present some CSP plant installed in Morocco and in the world, their capacities and the year of the construction. First of all, it is obvious that current CSP plants are operated just with the four geometries of solar concentrators: parabolic, trough parabolic, heliostats and Fresnel mirrors. The

Table 5.2 Efficiency and Applications for Parabolic Solar Concentrator PC [13, 77–79]

Parabolic Solar Concentrator PC				
Efficiency	– 17.6% with incident solar radiation of 725 W/m² – Efficiencies 24% and 50% respectively in desert and humid climates – 70% by application in a stand-alone system for off-grid electrification – 22% as a solar cooker			
Applications	– Electricity generation – Off-grid electrification – Hybridization and storage – Cooking – Water distillation and desalination – Water pumping and irrigation			
Overview of CSP plant	**Power Plant**	**Location**	**Capacity (MW)**	**Year**
	Tooele Army Depot	United States	1.5	N/A
	Maricopa	United States	1.5	2010

Table 5.3 Efficiency and Applications for Parabolic Trough Solar Concentrator PTC [13, 77–79]

Parabolic Trough Solar Concentrator PTC				
Efficiency	– 3% and 38% in electricity and thermal generation, respectively – 70% as an operating performance in sunny and cloudy weather conditions			
Applications	– Electricity generation – Seawater desalination – Cooling system – Water pasteurization process – Solar heat generation in industrial and low-enthalpy processes – Production of chilled water			
Overview of CSP plant	**Power Plant**	**Location**	**Capacity (MW)**	**Year**
	NOOR I	Morocco	160	2015
	NOOR II	Morocco	200	2018
	CSNP Urat	China	100	2020
	Enerstar	Spain	50	2013
	Genesis	United States	250	2014
	La Risca	Spain	50	2009
	Solar Electric Generating Station I	United States	13.8	1984

Table 5.4 Efficiency and Applications for Heliostats (Power Tower) PT [13, 77–79]

Heliostats (Power Tower) PT				
Operating temperature	600°C by using molten salts and water			
Applications	Electricity generation			
Overview of CSP plant	**Power Plant**	**Location**	**Capacity (MW)**	**Year**
	NOOR III	Morocco	150	2018
	Atacama I/ Cerro Dominador	Chile	110	2021
	CEEC Hami	China	50	2019
	CRS	Spain	5	2012
	Planta Solar 10	Spain	11	2007
	Solar Two	United States	10	1995
	Solar One	United States	10	1982

Table 5.5 Efficiency and Applications for Fresnel Mirrors FM [13, 77–79]

Fresnel Mirrors FM				
Efficiency	– 25% through its application as a supported system to generate electricity – 85% in electricity generation with combination with gas turbine – 75% in cooling generation			
Applications	– Main heat source – Cogeneration system – Power source for a double-acting absorption chiller			
Overview of CSP plant	**Power Plant**	**Location**	**Capacity (MW)**	**Year**
	Lanzhou Dacheng Dunhuang	China	50	2019
	eLLO	France	9	2019
	Dhursar	India	125	2014

capacities of the installed CSP plant varies between 1.5 MW and 250 MW (Tables 5.2, 5.3, 5.4 and 5.5). The other six geometries – elliptical hyperboloid concentrator, Fresnel lens solar concentrator, compound parabolic solar concentrator, total internal reflection dielectric concentrator, quantum dot concentrator and conical solar concentrator – have not yet been used on CSP (Table 5.6). In fact, electricity generation and water desalination are considered as the majority applications of solar concentrators with efficiencies varying between 3% and 70% for electricity (3%, 25% and 70% by using respectively PTC, FM and PC) and 50% for water desalination (by means of PC). Combined with other devices, Fresnel mirrors generate electricity with 85% by its combination with gas turbine. In addition, between 50% and 53%, Fresnel lens concentrators and compound parabolic concentrators generate electricity as integrated devices with photovoltaic modules. Other applications of solar concentrator technologies vary between irrigation, solar cooking, heat and cooling generation, solar controller and water pasteurization processes. In this context, parabolic trough solar concentrator and Fresnel mirrors produce heat and cooling respectively with efficiencies of 38% and 75%. As a solar cooker, the parabolic concentrator possesses a performance, which can reach 22 %.[13]

Since the 1980s, as mentioned earlier, the CSP plants are based at most on PC, PTC, PT and FM. Despite the performance of parabolic concentrator in desert and humid climates (respectively 24% and 50%), just two CSP plants were constructed in United States. While the parabolic trough concentrator generates electricity with an efficiency of 3%, the parabolic one reaches up to 70% as a main source in electrification process. Fresnel mirrors, whether they are the primary or secondary source, work with other devices (such as the gas turbine mentioned in the previous paragraph). It is just in the cooling generation process where Fresnel mirrors are used as a main power source with an efficiency of 75%. In major applications, on

Table 5.6 Efficiency and Applications for EHC, FLC, CPC, DC_TIR, QDC and CC [13, 77–79]

Solar Concentrator	*EHC*	*FLC*	*CPC*	*DC_TIR*	*QDC*	*CC*
Efficiency	– 50% in water desalination process	– 50% as an integrated device with photovoltaic modules to generate electricity	– 53% as an efficiency through the hybrid system photovoltaic/ Compound parabolic solar concentrator	N/A	N/A	N/A
Application	– water desalination	– Photovoltaics (Electricity generation) – Concentration of solar radiation – Condensation – Photoelectric – Photovoltaics – Solar control in buildings	– Combination of the compound solar parabolic concentrator with Fresnel mirrors for steam generation and water distillation – Heat source in water heater system with storage	– Secondary concentrator for space solar thermal applications at NASA	N/A	N/A
Not integrated in CSP Plant			• Elliptical hyperboloid concentrator: EHC • Fresnel lens solar concentrator: FLC • Compound parabolic solar concentrator: CPC • Dielectric concentrator with total internal reflection: DC_TIR • Quantum dot concentrator: QDC • Conical solar concentrator: CC			

the other hand, parabolic geometries work as the primary source alone. For the heliostats, the concentrated solar radiations through vaporize molten salts and water to reach up to 600°C for driving a turbine and generate electricity.[13, 77]

For the solar concentrator technologies not integrated in the CSP plant, they operate in several applications with efficiencies varying between 50% and 53%. As PTC, PT and FM (with the expectation of cooling generation) worked as supported sources, the FLC and CPC operate likewise as integrated sources with other devices. In addition, although the elliptical hyperboloid concentrators EHC is not integrated in CSP plants, they concentrate solar radiation to desalinate seawater to deal with the hydric stress trouble, with an efficiency which reaches up to 50%. The compound parabolic concentrator, in addition to generating electricity as a support system, produces heat as main source and in combination with Fresnel mirrors.[13, 78, 79]

What about developing the first and the second generations of CSP plants? Is it enough to drive CSP plants with other more efficient heat transfer fluids to introduce third-generation CSP plants? If it is enough, what about the optical performance of solar concentrator?

The analyses developed previously indicate that the non-integrated solar concentrator in CSP plant benefits of a high performance and efficiency which can reach up to 53%. Therefore, the application of the EHC, FLC and CPC as main or supported sources in CSP plants of first and second generations can lead to many satisfactions of energy needs: electrification, heat generation, cooking and seawater desalination. Regarding the performance of the parabolic trough solar concentrator in sunny and cloudy weather conditions (with 70% of efficiency), one of the three systems – EHC, FLC and CPC – can be combined with the PTC system to increase the electricity generation and the other applications efficiencies. Also, the parabolic solar concentrators must be applied even more in CSP plants, alone or by combining with other solar concentrators, to develop energy production and optimize new systems in solar cooking, desalination water and other applications. Concerning the three systems – dielectric concentrator with total internal reflection, quantum dot concentrator and conical solar concentrator – they need more development and research work to be used in CSP plants, either to optimize first and second generation or to develop new generation to eliminate the portion of fossil energy and make a sustainable energy consumption.

5.5 CONCLUSION

This chapter presents an overview and analyses the current solar concentrators technologies, parabolic solar concentrators, parabolic troughs, heliostats, Fresnel mirrors, Fresnel lens concentrators, elliptical hyperboloid concentrators, compound parabolic solar concentrators, dielectric concentrators with total internal reflection, quantum dot concentrators and conical solar concentrators and their role to develop the CSP plants. The analysis in this chapter is as follows:

- An introduction about the dilemma between renewable energy and fossil energy and the role played by renewable energy to make a sustainable world because fossil energy has dominated since the Industrial Revolution.
- A description of each solar concentrator system by giving their first emergence, applications and efficiencies.
- A classification of the systems into non-integrated and integrated solar concentrators in the CSP plant.
- A comparison between each solar concentrator type.

The main findings of this analysis were:

- Parabolic solar concentrators, parabolic trough solar concentrators, heliostats and Fresnel mirrors are the only concentrator systems integrated in the CSP plant.
- The CSP plants constructed by parabolic trough concentrators, heliostats and Fresnel mirrors are the most used compared to the CSP with parabolic concentrator.
- Parabolic solar concentrators generate electricity as main source with an efficiency of 70%. Its performance reaches 22% when is used as a solar cooker.
- Parabolic trough solar concentrator generates electricity and heat respectively with 3% and 38%.
- The operating temperature of heliostat receivers reaches 600°C by using molten salts and water as heat transfer fluid.
- Fresnel mirrors are applied as a main source to generate cooling with an efficiency of 75%. As a combined device with a gas turbine, it has a performance of 85% for Fresnel mirrors to produce electricity.
- Elliptical hyperboloid concentrators possess a performance of 50% in water desalination to deal with the hydric stress issue.
- Fresnel lens concentrators and compound parabolic concentrators operate as supported systems with a photovoltaic module to produce electricity with efficiencies varying from 50% to 53%.
- Dielectric concentrators with a total internal reflection, quantum dot concentrator and conical solar concentrator are not yet developed in the literature, neither are applied in electricity and heat generation nor in CSP plants.

According to these findings, concentrator solar technologies still require research work in order to optimize the energy (electricity and heat) production and to deal with several issues such as the hydric stress and the dominance of fossil fuels in the energy mix:

- The use of the elliptical hyperboloid concentrator, Fresnel lens concentrator and compound parabolic concentrator as main or supported sources in CSP plant to optimize the first and second generations.
- The combination of the elliptical hyperboloid concentrator, Fresnel lens concentrator and compound parabolic concentrator with the parabolic concentrator to increase the electricity generation.
- The combination of the parabolic concentrator with other solar concentrators to develop energy production and optimize new systems in solar cooking, desalination water and other applications.
- The development of dielectric concentrator with total internal reflection, quantum dot concentrator and conical solar concentrator to be used in CSP plants to optimize the first and second generation.

REFERENCES

[1] Peter Grajzl, Peter Murrell, Caselaw and England's economic performance during the Industrial Revolution: Data and evidence, *Journal of Comparative Economics*, Volume 52, Pages 145–165, Available online 2 November 2023.

[2] Mauro Rota, Luca Spinesi, Economic growth before the Industrial Revolution: Rural production and guilds in the European Little Divergence, *Economic Modelling*, Volume 130, January 2024, 106590.

[3] Fan Zhang, Shuzhong Wang, Yuanwang Duan, Wenjing Chen, Zicheng Li, Yanhui Li, Thermodynamic assessment of hydrothermal combustion assisted fossil fuel in-situ gasification in the context of sustainable development, *Fuel*, Volume 335, 1 March 2023, 127053.

[4] Alex O. Acheampong, Eric Evans Osei Opoku, Environmental degradation and economic growth: Investigating linkages and potential pathways, *Energy Economics*, Volume 123, July 2023, 106734.

[5] Abul Abrar Masrur Ahmed, Nadjem Bailek, Laith Abualigah, Kada Bouchouicha, Alban Kuriqi, Alireza Sharifi, Pooya Sareh, Abdullah Mohammad Ghazi Al khatib, Pradeep Mishra, Ilhami Colak, El-Sayed M. El-Kenawy, Global control of electrical supply: A variational mode decomposition-aided deep learning model for energy consumption prediction, *Energy Reports*, Volume 10, November 2023, Pages 2152–2165.

[6] Jiannan Wang, Waseem Azam, Natural resource scarcity, fossil fuel energy consumption, and total greenhouse gas emissions in top emitting countries, *Geoscience Frontiers*, Volume 15, Issue 2, March 2024, 101757.

[7] Chen Liu, Zhen Shao, Jianling Jiao, Shanlin Yang, How connected is withholding capacity to electricity, fossil fuel and carbon markets? Perspectives from a high renewable energy consumption economy, *Energy Policy*, Volume 185, February 2024, 113937.

[8] João Estevão, José Dias Lopes, SDG7 and renewable energy consumption: The influence of energy sources, *Technological Forecasting and Social Change*, Volume 198, January 2024, 123004.
[9] Hamed H. Pourasl, Reza Vatankhah Barenji, Vahid M. Khojastehnezhad, Solar energy status in the world: A comprehensive review, *Energy Reports*, Volume 10, November 2023, Pages 3474–3493.
[10] Md. Ibthisum Alam, Mashrur Muntasir Nuhash, Ananta Zihad, Taher Hasan Nakib, M. Monjurul Ehsan, Conventional and emerging CSP technologies and design modifications: Research status and recent advancements, *International Journal of Thermofluids*, Volume 20, November 2023, 100406.
[11] R. P. Praveen, Mohammad Abdul Baseer, Ahmed Bilal Awan, Muhammad Zubair, Performance analysis and optimization of a parabolic trough solar power plant in the middle east region, *Energies*, 2018, 11, 741.
[12] Guiqiang Li, Qingdong Xuan, M. W. Akram, Yousef Golizadeh Akhlaghi, Haowen Liu, Samson Shittu, Building integrated solar concentrating systems: A review, *Applied Energy*, Volume 260, 15 February 2020, 114288.
[13] Abdul Hai Alami, A. G. Olabi, Ayman Mdallal, Ahmed Rezk, Ali Radwan, Shek Mohammod Atiqure Rahman, Sheikh Khaleduzzaman Shah, Mohammad Ali Abdelkareem, Concentrating solar power (CSP) technologies: Status and analysis, *International Journal of Thermofluids*, Volume 18, May 2023, 100340.
[14] Evangelos Bellos, Christos Tzivanidis, Solar concentrating systems and applications in Greece – A critical review, *Journal of Cleaner Production*, Volume 272, 1 November 2020, 122855.
[15] Salvatore Guccione, Rafael Guedez, Techno-economic optimization of molten salt based CSP plants through integration of supercritical CO2 cycles and hybridization with PV and electric heaters, *Energy*, Volume 283, 15 November 2023, 128528.
[16] S. P. Singh, D. K. Gupta, Pradeep Yadav, Sachin Singh, Abhijeet Kumar, Influence of process parameters on capital cost, the efficiency of CSP based solar power plants- A review, *Materials Today: Proceedings*, Volume 62, Part 1, 2022, Pages 123–130.
[17] Charles-Alexis Asselineau, Armando Fontalvo, Shuang Wang, Felix Venn, John Pye, Joe Coventry, Techno-economic assessment of a numbering-up approach for a 100 MWe third generation sodium-salt CSP system, *Solar Energy*, Volume 263, October 2023, 111935.
[18] Yecheng Yao, Jing Ding, Shule Liu, Xiaolan Wei, Weilong Wang, Jianfeng Lu, Thermodynamic assessment of binary chloride salt material for heat transfer and storage applications in CSP system, *Solar Energy Materials and Solar Cells*, Volume 256, 1 July 2023, 112333.
[19] Chung-Yu Yeh, J. K. De Swart, Amirhoushang Mahmoudi, Abhishek K. Singh, Gerrit Brem, Mina Shahi, Simulation-based analysis of thermochemical heat storage feasibility in third-generation district heating systems: Case study of Enschede, Netherlands, *Renewable Energy*, Volume 221, February 2024, 119734.
[20] Yuanting Zhang, Yu Qiu, Qing Li, Asegun Henry, Optical-thermal-mechanical characteristics of an ultra-high-temperature graphite receiver designed for concentrating solar power, *Applied Energy*, Volume 307, 1 February 2022, 118228.

[21] Sajid Abbas, Yanping Yuan, Atazaz Hassan, Jinzhi Zhou, Ammar Ahmed, Li Yang, Emmanuel Bisengimana, Effect of the concentration ratio on the thermal performance of a conical cavity tube receiver for a solar parabolic dish concentrator system, *Applied Thermal Engineering*, Volume 227, 5 June 2023, 120403.
[22] Elnaz Bagherzadeh-Khajehmarjan, Seyyedeh Mahdieh Shakouri, Arash Nikniazi, Sohrab Ahmadi-Kandjani, Boosting the efficiency of luminescent solar concentrator devices based on CH3NH3PbBr3 perovskite quantum dots via geometrical parameter engineering and plasmonic coupling, *Organic Electronics*, Volume 109, October 2022, 106629.
[23] Arkbom Hailu Asfaw, Manual tracking for solar parabolic concentrator – For the case of solar injera baking, Ethiopia, *Heliyon*, Volume 9, Issue 1, January 2023, e12884.
[24] Keith Lovegrove, Wes Stein, Chapter 1 – Introduction to concentrating solar power technology, Concentrating Solar Power Technology (Second Edition), Principles, Developments, and Applications, *Woodhead Publishing Series in Energy*, 2021, Pages 3–17.
[25] Yasser Aldali, Abdelhamed Elmansuri, Basim Belgasim, Atul Sagade, Müslüm Arıcı, A simplified opto-thermal assessment and economic study of parabolic dish solar concentrator pondering on the receiver position induced uncertainties, *Thermal Science and Engineering Progress*, Volume 42, 1 July 2023, 101920.
[26] Katlego Lentswe, Ashmore Mawire, Prince Owusu, Adedamola Shobo, A review of parabolic solar cookers with thermal energy storage, *Heliyon*, Volume 7, Issue 10, October 2021, e08226.
[27] Hang-Suin Yang, Hao-Qiang Zhu, Xian-Zhong Xiao, Comparison of the dynamic characteristics and performance of beta-type Stirling engines operating with different driving mechanisms, *Energy*, Volume 275, 15 July 2023, 127535.
[28] Mohammad Amin Babazadeh, Mojtaba Babaelahi, Mahdi Saadatfar, Enhancing solar stirling engine performance through the use of innovative heat transfer fin shapes, *International Journal of Thermal Sciences*, Volume 190, August 2023, 108290.
[29] Sajjad A. Salih, Baseem A. Aljashaami, Naseer T. Alwan, Mohammed A. Qasim, Sergey E. Shcheklein, Vladimir I. Velkin, Isothermal thermodynamic analysis investigation of the Stirling engine types (alpha, beta, gamma): A theatrical study, *Materials Today: Proceedings*, https://doi.org/10.1016/j.matpr.2023.03.069, Available online 14 March 2023.
[30] Ammar S. Easa, Wael M. El-Maghlany, Mohamed M. Hassan, Mohamed T. Tolan, The performance of a gamma-type Stirling water dispenser with twin wavy plate heat exchangers, *Case Studies in Thermal Engineering*, Volume 39, November 2022, 102464.
[31] Khaled M. Bataineh, Moath F. Maqableh, A new numerical thermodynamic model for a beta-type Stirling engine with a rhombic drive, *Thermal Science and Engineering Progress*, Volume 28, 1 February 2022, 101071.
[32] Juan A. Auñon, José M. Pérez, María J. Martín, Fernando Auñon, Daniel Nuñez, Development and validation of a software application to analyze thermal and kinematic multimodels of Stirling engines, *Heliyon*, Available online 27 July 2023, e18487.

[33] Francesco Catapano, Carmela Perozziello, Bianca Maria Vaglieco, Heat transfer of a Stirling engine for waste heat recovery application from internal combustion engines, *Applied Thermal Engineering*, Volume 198, 5 November 2021, 117492.
[34] Jing-hu Gong, Junzhe Wang, Hu Xiaoli, Yong Li, Hu Jian, Jun Wang, Peter D. Lund, Cai-yun Gao, Optical, thermal and thermo-mechanical model for a larger-aperture parabolic trough concentrator system consisting of a novel flat secondary reflector and an improved absorber tube, *Solar Energy*, Volume 240, 1 July 2022, Pages 376–387.
[35] Punit V. Gharat, Snehal S. Bhalekar, Vishwanath H. Dalvi, Sudhir V. Panse, Suresh P. Deshmukh, Jyeshtharaj B. Joshi, Chronological development of innovations in reflector systems of parabolic trough solar collector (PTC) – A review, *Renewable and Sustainable Energy Reviews*, Volume 145, July 2021, 111002.
[36] V. K. Jebasingh, G. M. Joselin Herbert, A review of solar parabolic trough collector, *Renewable and Sustainable Energy Reviews*, Volume 54, February 2016, Pages 1085–1091.
[37] M. Allam, M. Tawfik, M. Bekheit, E. El-Negiry, Heat transfer enhancement in parabolic trough receivers using inserts: A review, *Sustainable Energy Technologies and Assessments*, Volume 48, December 2021, 101671.
[38] Ahmed Bilal Awan, M. N. Khan, Muhammad Zubair, Evangelos Bellos, Commercial parabolic trough CSP plants: Research trends and technological advancements, *Solar Energy*, Volume 211, 15 November 2020, Pages 1422–1458.
[39] X. Y. Tang, W. W. Yang, Y. Yang, Y. H. Jiao, T. Zhang, A design method for optimizing the secondary reflector of a parabolic trough solar concentrator to achieve uniform heat flux distribution, *Energy*, Volume 229, 15 August 2021, 120749.
[40] Gang Wang, Zhen Zhang, Tieliu Jiang, Jianqing Lin, Zeshao Chen, Thermodynamic and optical analyses of a novel solar CPVT system based on parabolic trough concentrator and nanofluid spectral filter, *Case Studies in Thermal Engineering*, Volume 33, May 2022, 101948.
[41] Tucker Farrell, Kidus Guye, Rebecca Mitchell, Guangdong Zhu, A non-intrusive optical approach to characterize heliostats in utility-scale power tower plants: Flight path generation/optimization of unmanned aerial systems, *Solar Energy*, Volume 225, 1 September 2021, Pages 784–801.
[42] Omar Behar, Abdallah Khellaf, Kamal Mohammedi, A review of studies on central receiver solar thermal power plants, *Renewable and Sustainable Energy Reviews*, Volume 23, July 2013, Pages 12–39.
[43] Gopalakrishnan Srilakshmi, V. Venkatesh, N. C. Thirumalai, N. S. Suresh, Challenges and opportunities for Solar Tower technology in India, *Renewable and Sustainable Energy Reviews*, Volume 45, May 2015, Pages 698–709.
[44] Francisco J. Collado, Jesus Guallar, Quick design of regular heliostat fields for commercial solar tower power plants, *Energy*, Volume 178, 1 July 2019, Pages 115–125.
[45] Arslan A. Rizvi, Syed N. Danish, Abdelrahman El-Leathy, Hany Al-Ansary, Dong Yang, A review and classification of layouts and optimization techniques used in design of heliostat fields in solar central receiver systems, *Solar Energy*, Volume 218, April 2021, Pages 296–311.

[46] Saeb M. Besarati, D. Yogi Goswami, Elias K. Stefanakos, Optimal heliostat aiming strategy for uniform distribution of heat flux on the receiver of a solar power tower plant, *Energy Conversion and Management*, Volume 84, August 2014, Pages 234–243.
[47] Mohammad Saghafifar, Kasra Mohammadi, Kody Powell, Design and analysis of a dual-receiver direct steam generator solar power tower plant with a flexible heliostat field, *Sustainable Energy Technologies and Assessments*, Volume 39, June 2020, 100698.
[48] https://patents.google.com/patent/DE326516C/en
[49] A. Barbón, N. Barbón, L. Bayón, J. A. Otero, Optimization of the length and position of the absorber tube in small-scale Linear Fresnel Concentrators, *Renewable Energy*, Volume 99, December 2016, Pages 986–995.
[50] A. E. Rungasamy, K. J. Craig, J. P. Meyer, Comparative performance evaluation of candidate receivers for an etendue-conserving compact linear Fresnel mirror field, *Solar Energy*, Volume 231, 1 January 2022, Pages 646–663.
[51] Evangelos Bellos, Progress in the design and the applications of linear Fresnel reflectors – A critical review, *Thermal Science and Engineering Progress*, Volume 10, May 2019, Pages 112–137.
[52] Abdelhamid Fadhel, Fathia Eddhibi, Kais Charfi, Moncef Balghouthi, Investigation of a Linear Fresnel solar collector (LFSC) prototype for phosphate drying, *Energy Nexus*, Volume 10, June 2023, 100188.
[53] P. K. Sen, K. Ashutosh, K. Bhuwanesh, Z. Engineer, S. Hegde, P. V. Sen, P. Davies, Linear Fresnel Mirror Solar Concentrator with Tracking, *Procedia Engineering*, Volume 56, 2013, Pages 613–618.
[54] Cesare Silvi, The Italian national solar energy history project, *Proceedings of ISES Solar World Congress 2007: Solar Energy and Human Settlement*, 3065–3068.
[55] K. S. Reddy, Tapas K. Mallick, T. Srihari Vikram, H. Sharon, Design and optimisation of elliptical hyperboloid concentrator with helical receiver, *Solar Energy*, Volume 108, October 2014, Pages 515–524.
[56] Visakh Chittalakkotte, Vinish Lazzar Vincent, Pradeep Valiyaparambil, Development of a solar energy based desalination system using a hyperboloid concentrator, *Materials Today: Proceedings*, Volume 46, Part 19, 2021, Pages 9771–9775.
[57] Daria Freier Raine, Firdaus Muhammad-Sukki, Roberto Ramirez-Iniguez, Tahseen Jafry, Carlos Gamio, Indoor performance analysis of genetically optimized circular rotational square hyperboloid (GOCRSH) concentrator, *Solar Energy*, Volume 221, June 2021, Pages 445–455.
[58] Imhamed M. Saleh Ali, T. Srihari Vikram, Tadhg S. O'Donovan, K. S. Reddy, Tapas K. Mallick, Design and experimental analysis of a static 3-D elliptical hyperboloid concentrator for process heat applications, *Solar Energy*, Volume 102, April 2014, Pages 257–266.
[59] Shen Liang, Xinglong Ma, Qian He, Zhenzhen Wang, Hongfei Zheng, Concentrating behavior of elastic Fresnel lens solar concentrator in tensile deformation caused zoom, *Renewable Energy*, Volume 209, June 2023, Pages 471–480.
[60] W. T. Xie, Y. J. Dai, R. Z. Wang, K. Sumathy, Concentrated solar energy applications using Fresnel lenses: A review, *Renewable and Sustainable Energy Reviews*, Volume 15, Issue 6, August 2011, Pages 2588–2606.

[61] Nicholas Yew Jin Tan, Xinquan Zhang, Dennis Wee Keong Neo, Rui Huang, Kui Liu, A. Senthil Kumar, A review of recent advances in fabrication of optical Fresnel lenses, *Journal of Manufacturing Processes*, Volume 71, November 2021, Pages 113–133.
[62] Xinglong Ma, Rihui Jin, Shen Liang, Shuli Liu, Hongfei Zheng, Analysis on an optimal transmittance of Fresnel lens as solar concentrator, *Solar Energy*, Volume 207, 1 September 2020, Pages 22–31.
[63] Chetan Y. Bachhav, Puskaraj D. Sonawwanay, Study on design and performance enhancement of Fresnel lens solar concentrator, *Materials Today: Proceedings*, Volume 56, Part 5, 2022, Pages 2873–2879.
[64] Xueyan Zhang, Shuoxun Jiang, Ziming Lin, Qinghua Gui, Fei Chen, Model construction and performance analysis for asymmetric compound parabolic concentrator with circular absorber, *Energy*, Volume 267, 15 March 2023, 126597.
[65] Rongji Xu, Yusen Ma, Meiyu Yan, Chao Zhang, Shuhui Xu, Ruixiang Wang, Effects of deformation of cylindrical compound parabolic concentrator (CPC) on concentration characteristics, *Solar Energy*, Volume 176, December 2018, Pages 73–86.
[66] Swati Arora, Harendra Pal Singh, Lovedeep Sahota, Manoj K. Arora, Ritik Arya, Sparsh Singh, Aayush Jain, Arvind Singh, Performance and cost analysis of photovoltaic thermal (PVT)-compound parabolic concentrator (CPC) collector integrated solar still using CNT-water based nanofluids, *Desalination*, Volume 495, 1 December 2020, 114595.
[67] Amit K. Singh, Praveen K. Srivastava, Akhoury S. K. Sinha, Gopal N. Tiwari, An estimation of bio-methane generation from photovoltaic thermal compound parabolic concentrator (PVT-CPC) integrated fixed dome biogas digester, *Biosystems Engineering*, Volume 227, March 2023, Pages 68–81.
[68] Ahed Hameed Jaaz, Husam Abdulrasool Hasan, Kamaruzzaman Sopian, Mohd Hafidz Bin Haji Ruslan, Saleem Hussain Zaidi, Design and development of compound parabolic concentrating for photovoltaic solar collector: Review, *Renewable and Sustainable Energy Reviews*, Volume 76, September 2017, Pages 1108–1121.
[69] Meng Tian, Yuehong Su, Hongfei Zheng, Gang Pei, Guiqiang Li, Saffa Riffat, A review on the recent research progress in the compound parabolic concentrator (CPC) for solar energy applications, *Renewable and Sustainable Energy Reviews*, Volume 82, Part 1, February 2018, Pages 1272–1296.
[70] Harjit Singh, Philip C. Eames, A review of natural convective heat transfer correlations in rectangular cross-section cavities and their potential applications to compound parabolic concentrating (CPC) solar collector cavities, *Applied Thermal Engineering*, Volume 31, Issues 14–15, October 2011, Pages 2186–2196.
[71] O. H. Cruz-Silva, O. A. Jaramillo, Mónica Borunda, Full analytical formulation for Dielectric Totally Internally Reflecting Concentrators designs and solar applications, *Renewable Energy*, Volume 101, February 2017, Pages 804–815.
[72] Firdaus Muhammad-Sukki, Siti Hawa Abu-Bakar, Roberto Ramirez-Iniguez, Scott G. McMeekin, Brian G. Stewart, Abu Bakar Munir, Siti Hajar Mohd Yasin, Ruzairi Abdul Rahim, Performance analysis of a mirror symmetrical dielectric totally internally reflecting concentrator for building integrated photovoltaic systems, *Applied Energy*, Volume 111, November 2013, Pages 288–299.

[73] Firdaus Muhammad-Sukki, Siti Hawa Abu-Bakar, Roberto Ramirez-Iniguez, Scott G. McMeekin, Brian G. Stewart, Nabin Sarmah, Tapas Kumar Mallick, Abu Bakar Munir, Siti Hajar Mohd Yasin, Ruzairi Abdul Rahim, Mirror symmetrical dielectric totally internally reflecting concentrator for building integrated photovoltaic systems, *Applied Energy*, Volume 113, January 2014, Pages 32–40.
[74] O. H. Cruz-Silva, O. A. Jaramillo, R. Castrejón-García, Heat transfer analysis of Dielectric Totally Internally Reflecting Concentrators transmitting concentrated solar energy, *Renewable Energy*, Volume 125, September 2018, Pages 55–63.
[75] A. J. Chattena, K. W. J. Barnhama, B. F. Buxtonb, N. J. Ekins-Daukesa, M. A. Malikc, A new approach to modelling quantum dot concentrators, *Solar Energy Materials & Solar Cells*, Volume 75, 2003, Pages 363–371.
[76] Mun Soo Na, Joon Yeal Hwang, Seong Geun Hwang, Joo Hee Lee, Gwi Hyun Lee, Design and Performance Analysis of Conical Solar Concentrator, *Journal of Biosystems Engineering,* Volume 43, Issue 1, March 2018, Pages 21–29. https://doi.org/10.5307/JBE.2018.43.1.021
[77] I. Arias, J. Cardemil, E. Zarza, L. Valenzuela, R. Escobar, Latest developments, assessments and research trends for next generation of concentrated solar power plants using liquid heat transfer fluids, *Renewable and Sustainable Energy Reviews*, Volume 168, October 2022, 112844.
[78] O. Achkari, A. El Fadar, Latest developments on TES and CSP technologies – Energy and environmental issues, applications and research trends, *Applied Thermal Engineering*, Volume 167, 25 February 2020, 114806.
[79] Ettore Morosini, Enrico Villa, Guglielmo Quadrio, Marco Binotti, Giampaolo Manzolini, Solar tower CSP plants with transcritical cycles based on CO2 mixtures: A sensitivity on storage and power block layouts, *Solar Energy*, Volume 262, 15 September 2023, 111777.

Chapter 6

Soft Computing Techniques in Wind Conversion Systems

A Review

Jyoti Bhattacharjee and Subhasis Roy

6.1 INTRODUCTION

Wind energy has gained popularity in recent decades since it is clean, renewable, sustainable, and cost-effective. Wind energy knowledge is essential for understanding its potential in a specific location, safeguarding the safety of wind energy conversion systems, and generating power via wind turbines [1]. Soft computing techniques play an important part in wind conversion systems, providing a sophisticated and flexible approach to addressing the challenges inherent in harvesting wind energy. Wind power has emerged as a key actor in the global energy scene as demand for sustainable and renewable energy sources grows [1, 2]. However, the unpredictable and dynamic nature of wind patterns offers considerable hurdles to the successful operation of wind conversion devices. Soft computing techniques, distinguished by their capacity to simulate and adapt to complex and uncertain systems, offer an invaluable option for improving the performance and dependability of wind energy systems. These approaches encompass fuzzy logic, neural networks, genetic algorithms, and swarm intelligence [3]. According to forecasts from the European Wind Energy Association, wind power investments are expected to remain steady at approximately €10 billion annually until 2015, with a growing emphasis on offshore projects. Projections indicate that wind energy investments in the EU-27 will escalate to around €20 billion by 2030, with 60% of these investments allocated to offshore assets. The safety and reliability of wind energy systems hinge on meticulous design, production, construction, seamless operation, and proper maintenance of their various components. Given that, a significant portion of the life cycle cost is associated with operational and maintenance activities, particularly for offshore farms in remote locations with challenging weather conditions [4–6]. This chapter aims to achieve the following objectives to provide a systematic classification of existing literature on soft computing techniques in wind conversion systems identify trends and propose future research directions in the realm of soft computing [4–7].

Soft computing techniques in wind energy conversion systems transcend conventional methods, pushing boundaries and exploring innovative

DOI: 10.1201/9781003462460-6

applications, such as utilizing a hybrid fuzzy-neural network controller to optimize the performance of a vertical axis wind turbine (VAWT) in an urban environment. [5]. Wind energy faces special hurdles in urban locations, where buildings and other structures influence complicated wind patterns. To dynamically alter the blade pitch and rotational speed of the VAWT, a hybrid controller combines the adaptability of fuzzy logic with the learning capabilities of a neural network. This maximizes energy extraction in the face of shifting urban wind conditions and minimizes the impact on nearby structures, demonstrating the versatility of soft computing techniques (SCTs) in atypical environments [6]. Offshore wind farms confront particular problems, such as unexpected weather patterns and changing water conditions. Traditional control systems struggle to adapt, resulting in inefficient energy extraction.

In contrast, fuzzy logic controllers were created to dynamically modify the pitch and yaw of turbines depending on real-time wind and sea condition data. The outcome significantly improved energy yield compared to fixed control solutions, demonstrating the adaptability and efficiency of soft computing techniques in demanding circumstances [7]. Neural networks can coordinate a network of geographically scattered wind farms, balancing energy supply and demand throughout the grid. This sophisticated choreography enables effective power distribution and unlocks the full potential of renewable energy. Soft computing's adaptability shows as wind farms grow and technologies improve. Algorithms are easily adaptable to new turbine configurations, grid integration issues, and environmental fluctuations, making them future-proof partners in the clean energy revolution. Wind energy is one of many renewable energy sources that are quickly expanding, as seen

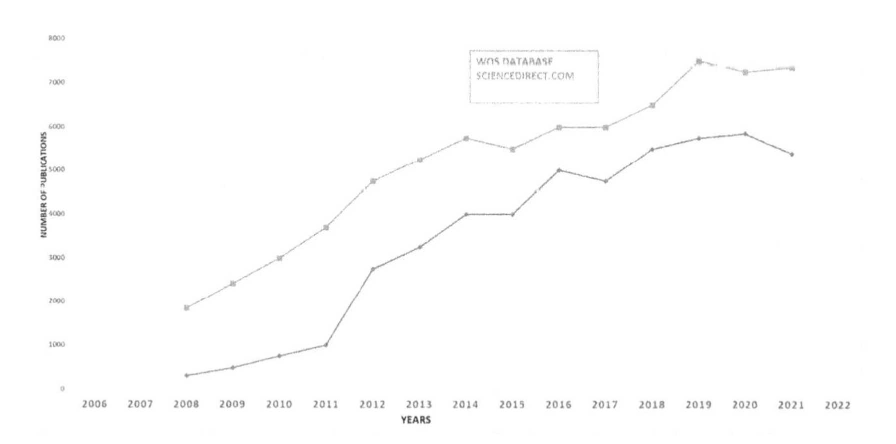

Figure 6.1 Number of publications based on soft computing techniques in wind energy. (Data input from WOS and ScienceDirect.com).

in. Figure 6.1 depicts the increase in papers on soft computing techniques in wind energy according to WOS and the sciencedirect.com database.

6.2 FUNDAMENTALS OF WIND ENERGY

Wind, a consequence of solar radiation differentials and rotation of the Earth, is a dynamic and ever-present energy source. Understanding the fundamental principles of wind entails recognizing the interaction between atmospheric conditions, terrain, and energy transfer from the sun. The kinetic energy of flowing air masses is a catalyst for a renewable energy source with enormous potential. Empirical mode decomposition (EEMD) combined with a genetic algorithm backpropagation-neural network (GABP-NN) was employed to predict short-term wind speed in an Inner Mongolia (China) wind farm. The devised model outperformed the conventional GA-BPNN method and introduced an enhanced error feedback scheme to forecast short-term wind speed at a wind farm in Taiwan [8]. Wind measurement and characterization are critical for building efficient wind energy conversion systems. Anemometers and wind vanes are traditional equipment for measuring wind speed and direction. Understanding wind patterns over time enables the selection of suitable locations for wind farms, ensuring that energy extraction is reliable and sustainable [9].

6.2.1 Fundamentals of Wind Energy Conversion Systems

This section describes the components of the wind energy system.

- Wind Turbines, From Traditional to Modern Designs: Wind turbines, the workhorses of wind energy conversion, have evolved from traditional windmills to sophisticated equipment optimized for optimal energy acquisition. A modern wind turbine's basic structure includes blades, a rotor, a nacelle, and a tower. The wind's kinetic energy of the wind rotates the blades, starting the conversion process. There are various designs, including horizontal and vertical axes [10].
- Rotor and Blades: The rotor, which consists of rotating blades, is the major element responsible for harnessing wind energy. Advances in blade design, including materials and aerodynamics, contribute to higher efficiency and energy capture.
- Nacelle: The nacelle, located atop the tower, houses critical components such as the generator, gearbox, and control systems. Nacelle design innovations aim to lower weight, increase dependability, and improve overall performance [11].

- Tower: The tower offers elevation, allowing the rotor to access higher wind speeds. Material strength, height optimization, and structural integrity are all essential factors to consider while designing a tower [12].
- The Generator: The generator is the core of the turbine, converting rotational energy into electricity. Electromagnetism dances within the generator, creating electrical current as magnets and coils contact.
- The Grid Connection: Before being injected into the grid to supply residential and commercial sectors, the generated electricity undergoes processing and voltage augmentation through transformers [13].
- Horizontal Axis Wind Turbines (HAWT): This is the most prevalent type, with its characteristic blades spanning the horizon. HAWTs are available in two- and three-bladed versions and are best suited for places with consistent, high-speed winds [14].
- Vertical Axis Wind Turbines (VAWT): These unusual dancers spin their blades on a vertical axis, providing advantages in urban areas and difficult wind regimes. Due to their modest size, they are ideal for rooftop installations. Figure 6.2 depicts the entire wind conversion system and the classification based on turbine technology, application type, and rotational speed.

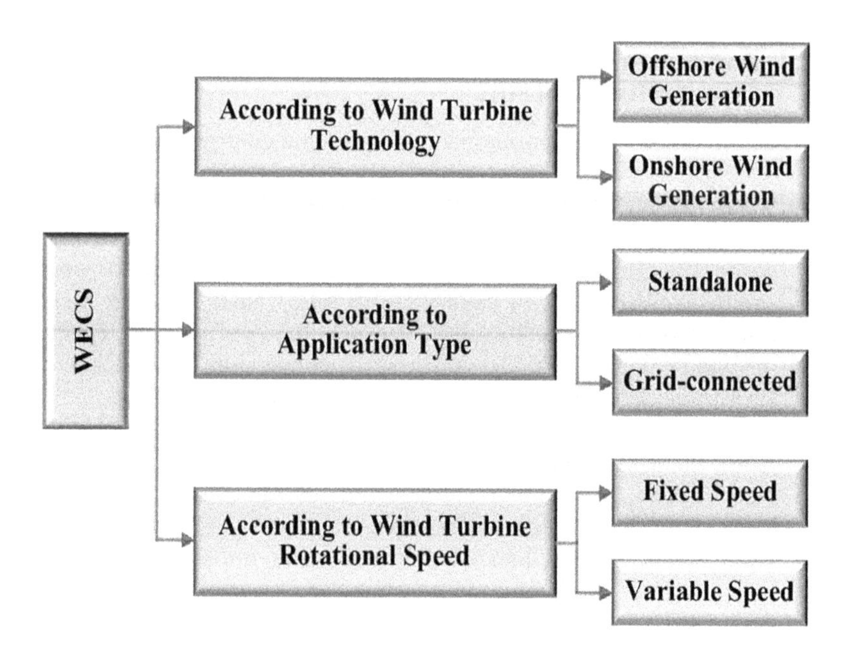

Figure 6.2 Classification of wind energy conversion system [15]. (Adapted from Springer 2022).

Wind energy conversion systems (WECS) are divided into two categories depending on the rotational speed of the wind turbine: variable speed and fixed speed. Fixed-speed systems have long been predominant due to their cost-effectiveness, simplicity, and robust reliability. They held sway until approximately two decades ago when variable speed systems started to garner attention. This transition can be attributed to the inherent constraints of fixed-speed systems, encompassing rigidity in adapting to grid voltage variations, apprehensions about mechanical stress, diminished wind energy conversion efficiency, and inevitable power flicker. Additionally, any variability in wind power instantaneously affects the grid [16]. Variable speed systems are increasingly gaining popularity, offering several advantages such as swift responses under transient power system conditions, enhanced wind energy conversion efficiency, and improved power output. Additionally, the use of a high-pole number electric machine may eliminate the need for a gearbox. However, it comes with drawbacks, including higher losses from additional power electronic equipment, increased costs, and added control complexity.

6.2.2 Power Generation

This chapter explores the synergy between power generation from wind energy and the transformative impact of soft computing procedures, unraveling their potential through examining existing literature and advancements in the field. Wind energy conversion technologies, most typically exhibited in wind turbines, have evolved dramatically. The movement of turbine blades, which drive generators, converts the wind's kinetic energy into electrical power in these systems. Wind power is widely adopted as an environmentally benign and renewable energy source worldwide. However, the fluctuation of wind conditions creates problems with the efficiency and dependability of standard wind turbines, necessitating creative ways for efficient power generation. Fuzzy logic, neural networks, and evolutionary algorithms are fundamental components of soft computing that contribute to improving wind power generation [17]. Fuzzy logic controllers were used in a wind farm environment to control the pitch and yaw of turbines based on real-time wind speed and direction data. Compared to fixed control systems, this resulted in a considerable increase in energy yield, demonstrating the efficacy of fuzzy logic in optimizing power generation [18].

Inspired by the human brain, neural networks excel at pattern detection and data processing. In wind power generation, neural networks are used for predictive maintenance, recognizing potential flaws before they cause system failures. For instance, a wind farm used a neural network trained on historical data to forecast approaching gearbox faults. The technology successfully discovered patterns suggesting possible defects, enabling preventative maintenance and saving costly breakdowns. This demonstrates how neural networks contribute to wind turbine reliability and longevity [19]. Natural selection-inspired evolutionary algorithms present a novel way to optimize wind energy systems. These algorithms use iterative procedures to fine-tune characteristics such as turbine blade designs and control systems, simulating the evolutionary process to maximize energy output [20].

6.2.2.1 Advantages and Issues of Soft Computing Techniques in Wind Power Generation

This section dives into the broad benefits and issues related to integrating soft computing techniques in wind power generation.

Advantages:

- Improved Efficiency: SCTs offer adaptive control systems, ensuring turbines operate efficiently in fluctuating environmental circumstances.
- Predictive Maintenance: Neural networks help with predictive maintenance, reducing downtime and prolonging wind turbine lifespan [20].
- Optimal Resource Utilization: The adaptability of SCTs provides optimal resource utilization, leading to consistent and reliable energy output.
- Environmental Impact Mitigation: Evolutionary algorithms make designing turbine designs that limit noise pollution and bird collisions easier, hence minimizing environmental impact [21].

Challenges:

- Computational Complexity: Putting SCTs into practice may provide computational hurdles, necessitating sturdy hardware and efficient algorithms.
- Data Dependence: SCTs' efficiency depends on having access to extensive and reliable data, which can be difficult in data-scarce contexts.
- Interpretability: Due to the intrinsic complexity of SCTs, interpreting decision-making processes may be difficult, posing problems in crucial applications [22]. Soft computing techniques and their application in wind power generation are positioned to grow further as they evolve, and the integration of soft computing techniques in wind power generation signifies a dramatic move towards flexible, efficient, and sustainable energy solutions.

In contemporary times, the conversion of variable frequency and voltage on the generator side to fixed frequency and voltage on the grid side can be achieved without the need for cumbersome and short-lived DC-link devices. Matrix converters facilitate the direct conversion of AC/AC power in a single stage, leading to exceptional efficiency and a compact footprint. Moreover, matrix converters emerge as a reliable choice for wind turbines, especially in offshore applications, as they enable the four-quadrant operation and can be seamlessly employed with both synchronous and induction wind turbines [23].

6.3 INTRODUCTION TO SOFT COMPUTING

Soft computing, distinguished by its capacity to deal with uncertainty and imprecision, has seen a significant transition in recent years [24]. Integrating neural networks, evolutionary algorithms, fuzzy logic, and swarm intelligence, among other techniques, has opened up new horizons of computer intelligence.

6.3.1 Fuzzy Logic

Fuzzy logic becomes useful for nonlinear systems. The wind turbine system comprises various nonlinearities that must be considered during modeling. Maximum power searches, pitch angle control, and turbine rotor speed are some applications that can benefit from fuzzy logic controllers and fuzzy inference systems. The fuzzy rules represent the knowledge and abilities of a human operator who makes the necessary modifications to manage the system with minimal mistakes and rapid reaction [25]. Fuzzy logic controllers (FLCs) are extremely useful in adaptive control systems when accurate mathematical modeling is difficult. An example shows how FLCs can be used to regulate an inverted pendulum, demonstrating their capacity to adapt to dynamic and uncertain contexts. Owing to their versatility, fuzzy logic has emerged as a significant role in systems where standard control methods fall short.

Real-world decision-making frequently involves unclear and inaccurate data. Fuzzy decision-making models provide a structured approach to dealing with ambiguity, and their applications range from company strategy to medical diagnostics [26]. Combining neural networks with fuzzy logic creates an enticing synergy that capitalizes on the strengths of both paradigms. The variety of fuzzy logic concepts and neural network topologies results in neuro-fuzzy systems with improved adaptability, learning capabilities, and interpretability.

Neuro-fuzzy systems combine neural network learning capabilities with the interpretability of fuzzy logic. This hybrid technique has been used in various fields, including control systems, forecasting, and pattern recognition. An example demonstrates the potential of a neuro-fuzzy system to adapt to changing environmental conditions by predicting energy usage in smart buildings [27]. In real-world applications, the interpretability of AI models is crucial. The combination of fuzzy logic and neural networks aids in the construction of explainable AI models. In one case, an answerable neuro-fuzzy system is used in medicine, where interpretability is critical for establishing trust and acceptance [28]. A study showed researchers achieved a 97% accuracy in wind speed forecasting using a hybrid FTS-ANN (artificial neural network) model, demonstrating its effectiveness in anticipating wind patterns. Combining fuzzy logic with evolutionary algorithms like particle swarm optimization (PSO) or genetic algorithms (GA) can further enhance MPPT performance. For instance, researchers demonstrated a fuzzy-PSO-based MPPT controller that outperformed conventional MPPT techniques, leading to greater energy capture [28]. Figure 6.3 depicts the modeling of doubly fed induction generator (DFIG)-based WECS on the principle of fuzzy networks.

6.3.2 Neural Networks

The progression of soft computing methodologies is marked by a continual quest for intelligent solutions that emulate human-like cognitive processes. While traditional approaches remain valuable, they are being outpaced by remarkable advancements in neural networks and other groundbreaking paradigms within the realm of soft computing. This chapter aims to provide a comprehensive overview, analyze the current state-of-the-art technology, and shed light on the enormous potential offered by these cutting-edge technologies. Unlike their predecessors, these networks exhibit amazing resilience, learning capacities, and the ability to interpret complicated, nonlinear interactions [30]. This section goes into the principles of neural networks, emphasizing their structure, training procedures, and the wide variety of topologies that have arisen in recent years. Many neural network topologies are built on multilayer perceptrons (MLPs), which feature hidden layers capable of capturing subtle patterns, and have found applications ranging from image identification to financial forecasting. Recent advances in training methods and regularization approaches have improved the performance and interpretability of MLPs, contributing to their widespread

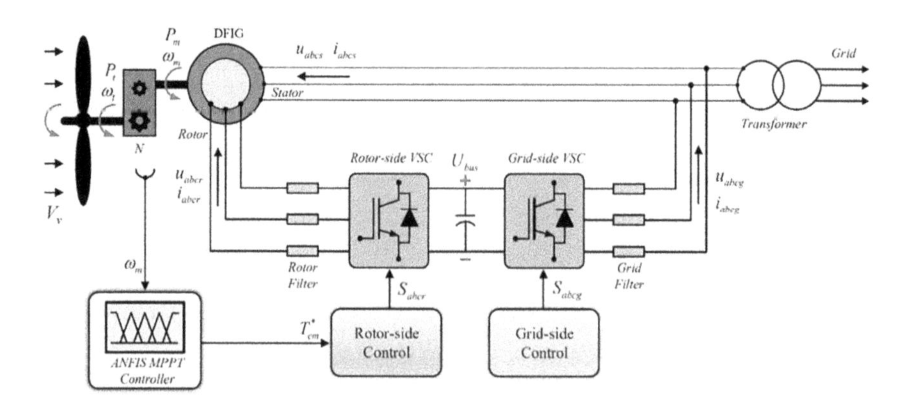

Figure 6.3 Grid-connected neuro-fuzzy system WECS configuration [29]. (Adapted from Elsevier 2014).

adoption [31]. Convolutional neural networks (CNNs) have redefined the accuracy and efficiency benchmarks in image and signal processing. This section delves into the architecture of CNNs, emphasizing their capacity to automatically learn hierarchical features, which makes them useful in applications such as picture categorization, object identification, and facial recognition. Case studies demonstrate the influence of CNNs in real-world applications [32]. The predictive potential of neural networks has been exploited in the financial sector for automated trading, risk assessment, and fraud detection. A study in India used ANNs to predict wind speed with 98% accuracy, enabling grid operators to make informed decisions about power dispatch and reserve allocation.

A neural network-based adaptive control system was used in an onshore wind farm in a mountainous terrain. The system constantly monitored wind speed, direction, and turbulence, altering turbine pitch and yaw to maximize energy harvest. As a result, overall energy yield increased significantly, demonstrating the versatility and efficiency of neural network-driven control systems. A neural network-based predictive maintenance system was developed at a wind farm in Northern Europe. The technology accurately anticipated gearbox failures weeks in advance by analyzing data on turbine vibrations, temperature fluctuations, and power output. This proactive strategy saved maintenance costs and avoided unscheduled shutdowns, demonstrating the usefulness of neural network-based systems in improving reliability [33]. A neural network-based wind forecasting system was deployed in an offshore wind farm as part of a research study. The neural network effectively forecasted wind patterns up to 48 hours in advance using historical weather data, sea conditions, and atmospheric pressure. This higher forecasting precision enabled better grid management and boosted overall energy efficiency. While incorporating neural networks into wind energy conversion systems has enormous promise, issues such as data scarcity, model interpretability, and computational complexity must be solved. Ongoing research aims to overcome these obstacles and further optimize the synergy between neural networks and wind energy technology [34]. The comprehensive control of machine-side and grid-side converters is investigated to extract maximum power from wind turbines and improve distribution grid power quality. For the forecast of wind speed, a suitable topology and architecture of an artificial neural network are utilized, which may be used to estimate power from many turbines, as required in the design of the wind-solar hybrid system. The goal was to increase design accuracy while lowering maintenance costs and losses [35].

6.3.3 Genetic Algorithms

Genetic algorithms (GAs) are a subset of soft computing techniques that imitate natural selection and evolution principles. This section discusses how GAs, with their potential to evolve solutions across generations, alter the landscape of wind energy conversion. Wind conditions are inherently changing, necessitating the dynamic adjustment of wind turbine control systems for best performance. In reaction to real-time environmental data, genetic algorithms can optimize control parameters such as pitch and yaw angles. This versatility guarantees that turbines function efficiently even when wind conditions change.Genetic algorithms were employed to optimize the control plan for a group of turbines in an offshore wind farm case study. The algorithm considered wind direction, sea status, and energy demand. Compared to fixed control systems, the evolved control strategy displayed a considerable improvement in energy yield, demonstrating the ability of GAs to adapt to complex and dynamic situations [36]. Traditional wind turbine designs frequently require trade-offs between efficiency, noise, and environmental impact. Genetic algorithms provide an innovative method for optimizing turbine blade designs, lengths, and configurations. By storing these characteristics in the chromosomal representation, GAs can evolve designs that maximize energy collection efficiency while minimizing noise and environmental damage. This versatility enables turbines to operate optimally under a variety of wind situations. Researchers used genetic algorithms to optimize the design of wind turbine blades in an onshore wind farm. The system considered various characteristics, including wind speed, turbulence, and noise levels. Compared to standard designs, the evolved blade designs displayed a considerable increase in energy capture efficiency and reduced noise levels [37]. The configuration of wind turbines inside a wind farm considerably impacts total energy collection efficiency. Wind direction, wake effects, and inter-turbine interactions can all be considered when using genetic algorithms to optimize turbine architecture. This optimization procedure guarantees that each turbine operates in the most favorable conditions, maximizing the total energy output of the wind farm. The researchers used genetic algorithms to optimize the configuration of turbines in an onshore wind farm. The algorithm took into account geography, wind patterns, and land limits. The modified wind farm architecture significantly improved energy capture efficiency, demonstrating the potential of GAs in optimizing large-scale wind energy systems [38]. While genetic algorithms provide tremendous benefits, their implementation in wind energy conversion is not without problems. These include computational intensity, convergence speed, and the necessity for precise modeling. Addressing these issues is critical for the successful implementation of GAs in real-world circumstances, A few challenges are highlighted here:

- Computational Intensity: The optimization process in genetic algorithms entails analyzing a large number of candidate solutions, which can be computationally costly. Efficient algorithms and parallel computing approaches are critical for reducing the computational burden, particularly in large-scale wind energy systems.

- Convergence Speed: The convergence speed of GAs hinges on parameters like solution representation, genetic operators, and population size. Striking a balance between exploration and exploitation is crucial to expedite the convergence toward optimal solutions. [39].

One of the strengths of GAs is their capacity to investigate many solutions concurrently. In wind energy conversion, this parallel development can optimize many factors simultaneously, such as turbine design, control strategies, and wind farm layout. The holistic approach of GAs adds to a more comprehensive optimization of the entire system. Future research avenues could include the creation of hybrid algorithms that integrate GAs with other soft computing techniques and the development of adaptive algorithms that modify their parameters dynamically during optimization. Illustrated in Figure 6.4 is the correlation between the tip-speed ratio and the power coefficient Cp of the wind utilizing genetic algorithms. Consequently, it can be asserted that determining a specific turbine rotational speed for a given wind velocity allows the extraction of the maximum achievable mechanical power from the wind [40].

6.3.4 Swarm Intelligence

Swarm intelligence (SI) emerges as a light of innovation. In contrast to conventional approaches, SI harnesses the collective intelligence of decentralized and self-organizing systems, taking inspiration from the cooperative

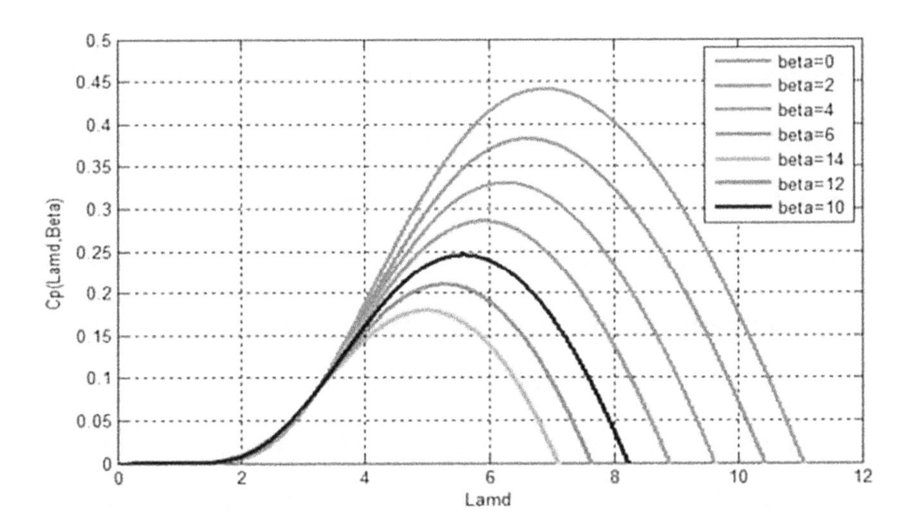

Figure 6.4 Graphical presentation of modeling of a wind power system using the genetic algorithm [40]. Reproduced with permission MDPI, © 2023.

behaviors observed materials in nature [41, 42]. Swarm intelligence, a subclass of soft computing, is inspired by the collective behaviors of social insects, birds, and other species. These decentralized systems exhibit extraordinary qualities such as self-organization, flexibility, and robustness in changing external conditions. SI concepts have been applied in various sectors, and their incorporation into wind energy conversion systems represents a paradigm shift toward efficiency and sustainability. Particle swarm optimization, a cornerstone of SI, has gained attention for its capacity to optimize complicated systems through the collective movement of particles in a search space. PSO has been used in wind energy to optimize turbine configuration in wind farms. PSO maximizes energy capture and minimizes wake effects by dynamically altering turbine positions based on real-time wind data, demonstrating a significant improvement over typically fixed configurations [42]. PSO was used to optimize turbine placements in a wind farm situated in a difficult landscape with changing wind patterns. The program altered the arrangement to leverage prevailing wind directions, minimizing turbulence and improving overall farm efficiency [43].

Ant colony optimization (ACO), inspired by ant foraging behavior, is another SI technique that has found application in wind energy optimization. ACO has been used in designing wind turbine blades to discover the ideal form while keeping aerodynamic efficiency and structural restrictions in mind. The decentralized structure of ACO allows for a thorough study of design alternatives, resulting in creative and efficient blade arrangements. In one review on blade design, ACO was used to explore a large design space while considering characteristics like lift, drag, and material restrictions. The algorithm converged on a unique blade design with higher aerodynamic performance, resulting in increased energy capture and lower structural loads [44]. Researchers are using a deep neural network to achieve just that, boosting energy output by 12% compared to traditional methods. Now wind farms can be arranged like synchronized dancers, capturing the most energy from the unpredictable movements of the wind. Maintaining the health of wind turbines is critical for minimizing downtime and guaranteeing steady electricity flow. Swarm intelligence methods such as glowworm swarm optimization (GSO) come into play here. GSO algorithms, inspired by glowworms' luminous communication, analyze sensor data from turbines, looking for patterns that suggest probable defects. Early problem identification, facilitated by GSO, can lead to speedier maintenance and lower repair costs [45].

Inspired by honeybee foraging behavior, the bee algorithm (BA) is a soft computing technique that excels at optimization tasks. BA has been used in wind energy for dynamic load balancing in wind farms. The program mimics collaborative bee communication and redistributes load across turbines in response to changing wind conditions, ensuring optimal energy harvest while minimizing the burden on individual turbines. BA was used to dynamically shift load among turbines in a wind farm subjected to fluctuating wind speeds and directions. This method ensured that each turbine functioned within its optimal range, reducing overloading and increasing the lifespan of the wind farm.

One of the primary benefits of swarm intelligence is its natural flexibility in dynamic and changing surroundings. Wind patterns are notoriously changeable, yet soft computing algorithms such as PSO and ACO excel at responding to real-time changes, ensuring that wind turbines function at peak efficiency regardless of fluctuations. Swarm intelligence algorithms are meant to explore global solution spaces while avoiding local optima. This property is especially useful in wind energy conversion, where determining the global optimum for characteristics such as turbine layout or control settings is critical for maximizing energy yield [46].

6.4 FUTURE TRENDS AND CHALLENGES

Contemporary wind energy conversion systems frequently operate as variable-speed technologies, aiming to optimize wind power generation by maintaining energy production below the rated power while minimizing stress on drivetrains. Over the past few decades, dedicated efforts from researchers and manufacturers have resulted in the introduction of various improved WECS technologies. This has led to a global enhancement in power production, heightened power reliability and quality, and reduced costs associated with wind energy. In this context, soft computing-based wind energy conversion technology emerges as the dominant system in both onshore and offshore wind energy industries, boasting an exceptionally attractive combination of high power production per cost performance [47, 48].

Future developments and challenges linked to the exclusive use of soft computing techniques (SCTs) in wind energy conversion systems are discussed in this chapter. From the hopeful trends transforming the environment to the complicated challenges that necessitate novel solutions, we begin on a tour through the evolving domain of SCTs in wind generation. Soft computing techniques have drastically altered the trajectory of wind energy conversion, enabling adaptive and intelligent solutions.

6.4.1 Advanced Machine Learning Models

The growth of machine learning models is an important trend in SCTs for wind energy conversion. Deep learning architectures and advanced neural networks optimize turbine management, predictive maintenance, and power forecasting. These algorithms use massive datasets to find detailed patterns, enabling more precise and sophisticated decision-making.

A recent study used a deep neural network model to anticipate wind power output with exceptional accuracy. The model considered complicated connections between climatic variables, historical power generation, and geographical features. The results showed a considerable improvement over traditional forecasting approaches, demonstrating the potential of sophisticated [49].

6.4.2 Hybrid Soft Computing Systems

Incorporating numerous SCTs into hybrid systems is gaining traction. Fuzzy logic controllers, neural networks, and evolutionary algorithms are being coupled to develop synergistic solutions that capitalize on the strengths of each technique. These hybrid systems seek to solve the limits of individual SCTs by providing more robust and adaptive control techniques. In a wind farm situation, a hybrid control system was designed that combined fuzzy logic for real-time adaptation, neural networks for predictive maintenance, and evolutionary algorithms for optimal turbine parameter tuning. This integrated strategy boosted energy collection efficiency, reduced maintenance costs, and increased system reliability [50].

6.4.3 Multi-Objective Optimization for Wind Farm Layout

Traditional wind farm layout design frequently focuses on maximizing individual turbine efficiency. However, future trends lead to multi-objective optimization employing evolutionary algorithms. These algorithms consider wake interference between turbines, environmental impact, and land use efficiency, resulting in more holistic and sustainable wind farm designs. In one instance, an evolutionary algorithm was used to optimize the layout of an offshore wind farm. The program considered energy generation, wake impacts, and seabed conditions all at once, resulting in a configuration that maximized energy yield while minimizing environmental damage [51].

- *Challenges:*
  While the future contains immense potential, it is not without obstacles. The sole reliance on soft computing techniques for wind energy conversion presents challenges that must be overcome to ensure computational power and cost; implementing complex soft computing techniques can be computationally expensive, particularly for edge computing applications. Optimizing algorithms and hardware efficiency will be critical to keeping the conductor's baton moving without breaking the bank. As wind energy infrastructure becomes more networked and reliant on software, sophisticated cybersecurity measures will be required to protect against cyberattacks that could

disrupt the flow of clean energy. Imagine the wind's whisper being hijacked, transforming the symphony into a cacophony of discord. The general adoption and success of these technologies.

- Data Dependency and Quality: SCTs' efficacy strongly depends on the availability and quality of data. As wind energy systems become increasingly complicated and linked, the demand for high-quality, diversified datasets becomes critical. Collecting and preserving such datasets poses challenges, particularly in remote or offshore sites. Machine learning algorithms thrive on historical data for training and optimization. The performance of these algorithms may be affected in areas with inadequate historical wind data. A wind farm in a region with limited historical wind data used satellite-based wind observations to supplement its dataset. Machine learning models were trained using ground-based and satellite data, displaying increased forecasting skills even in locations with low historical records.
- Real-Time Data Challenges: Many SCTs rely on real-time data for adaptive decision-making. Ensuring a robust and reliable real-time data infrastructure presents problems, particularly in offshore areas where connectivity and data transmission might be difficult. Due to the remote position, an offshore wind farm experienced problems with real-time data transmission. A hybrid strategy, combining on-site data storage with periodic data transmission, was used to overcome this. This method preserved the integrity of real-time data inputs while overcoming the obstacles of continuous connectivity [51, 52].
- The Neural Network Black-Box Challenge: Due to their black-box nature, neural networks offer interpretability issues. Understanding the decisions made by neural network-based controllers or predictive maintenance systems is critical for operators and stakeholders. For instance, a wind farm operator adopted a neural network-based controller but encountered pushback from maintenance crews due to a lack of transparency in decision-making. An explainability module was added to solve this, which provides insight into how the neural network selected control modifications. This increased trust and acceptance among the operational workforce [53–58].
- Computational Power and Cost: Implementing complex soft computing techniques can be computationally expensive, particularly for edge computing applications. As wind energy infrastructure becomes more networked and reliant on software, sophisticated cybersecurity measures will be required to protect against cyberattacks that could disrupt the flow of clean energy. The convergence of machine learning, evolutionary algorithms, and real-world applications showcases the immense potential of SCTs in optimizing energy capture, enhancing adaptability, and contributing to a sustainable future [59–61]. The histogram depicts the overall observed and estimated W values at the

testing stations. As can be observed, all of the models underestimate W for certain months while overestimating it for others. For instance, June, July, and August have the highest levels of W underestimate [61]. The performance of soft computing models along with their wind-speed has been depicted yearwise in Figure 6.5. The maximum wind speed has been found around the month of July and August.

6.5 CONCLUDING REMARKS

This chapter reviews the realms of fuzzy logic, neural networks, and evolutionary algorithms, revealing a landscape rich with possibilities, challenges, and transformative potential of wind energy conservation. The exploration of wind energy conversion system configurations, electrical generators, and diverse power converter topologies is complemented by the consideration of global grid standards encompassing codes related to frequency, power factor, power quality, and harmonics. The reliance on high-quality, real-time data, as well as frequently complex and non-intuitive outputs of the algorithms, pose significant obstacles. However, obstacles are not roadblocks, but rather chances for innovation. The appeal is for collaborative efforts that combine the strengths of SCTs with other emerging technologies. Hybrid models that combine physics-based techniques with the adaptability of SCTs give a possible option. These hybrid techniques have the potential to overcome the restrictions of data reliance and interpretability, providing a compass for future research endeavors. This convergence of technology and nature within holds the potential of a future in which wind energy, captured and optimized through the intelligence of soft computing techniques, becomes an intrinsic component of a cleaner, greener, and more sustainable world.

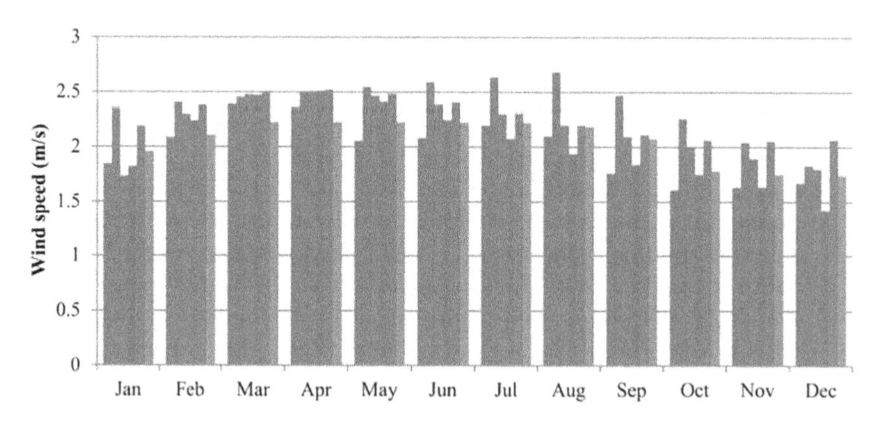

Figure 6.5 Overall analysis of wind speed based on the performance of soft computing models [61]. (Adapted from Elsevier 2022).

ACKNOWLEDGMENT

Author (S. Roy) would like to acknowledge 'Scheme for Transformational and Advanced Research in Sciences (STARS)' (MoE-STARS/STARS-2/2023–0175) by the Ministry of Education, Govt. of India for promoting translational India-centric research in sciences implemented and managed by Indian Institute of Science (IISc), Bangalore, for their support.

REFERENCES

[1] M. Abdelateef Mostafa, E. A. El-Hay, and M. M. ELkholy, "Recent trends in wind energy conversion system with grid integration based on soft computing methods: Comprehensive review, comparisons and insights," *Archives of Computational Methods in Engineering*, vol. 30, no. 3, pp. 1439–1478, Springer Science and Business Media LLC, 2022. https://doi.org/10.1007/s11831-022-09842-4

[2] S. S. Band, S. Ardabili, A. Mosavi, C. Jun, H. Khoshkam, and M. Moslehpour, "Feasibility of soft computing techniques for estimating the long-term mean monthly wind speed," *Energy Reports*, vol. 8, pp. 638–648, Elsevier BV, 2022. https://doi.org/10.1016/j.egyr.2021.11.247

[3] Z. Guo, W. Zhao, H. Lu, and J. Wang, "Multi-step forecasting for wind speed using a modified EMD-based artificial neural network model," *Renewable Energy*, vol. 37, no. 1, pp. 241–249, Elsevier BV, 2012. https://doi.org/10.1016/j.renene.2011.06.023

[4] O. Khan et al., "Comparative study of soft computing and metaheuristic models in developing reduced exhaust emission characteristics for diesel engine fueled with various blends of biodiesel and metallic nanoadditive mixtures: An ANFIS – GA – HSA approach," *ACS Omega*, vol. 8, no. 8, pp. 7344–7367, American Chemical Society (ACS), 2023. https://doi.org/10.1021/acsomega.2c05246

[5] Y. Ding et al., "Three-dimensionally ordered macroporous materials for photo/electrocatalytic sustainable energy conversion, solar cell and energy storage," *EnergyChem*, vol. 4, no. 4, p. 100081, Elsevier BV, 2022. https://doi.org/10.1016/j.enchem.2022.100081

[6] L. Naderloo, H. Javadikia, and M. Mostafaei, "Modeling the energy ratio and productivity of biodiesel with different reactor dimensions and ultrasonic power using ANFIS," *Renewable Sustainable Energy Review*, vol. 70, pp. 56–64, 2017. https://doi.org/10.1016/j.rser.2016.11.035

[7] J. Zhu, R. Li, Z. Wang, S. Liu, and H. Lv, "Decoupled analysis of the effect of hydroxyl functional groups on delay of ignition with fictitious hydroxyl," *Process Safety and Environmental Protection*, vol. 161, pp. 285–294, Elsevier BV, 2022. https://doi.org/10.1016/j.psep.2022.03.028

[8] A. B. Fadhil and L. I. Saeed, "Sulfonated tea waste: A low-cost adsorbent for purification of biodiesel," *International Journal of Green Energy*, vol. 13, no. 1, pp. 110–118, Informa UK Limited, 2015. https://doi.org/10.1080/15435075.2014.896801

[9] A. B. Fadhil, S. H. Sedeeq, and N. M. T. Al-Layla, "Transesterification of non-edible seed oil for biodiesel production: Characterization and analysis of biodiesel," *Energy Sources, Part A: Recovery, Utilization, and Environmental*

Effects, vol. 41, no. 7, pp. 892–901, Informa UK Limited, 2018. https://doi.org/10.1080/15567036.2018.1520367

[10] A. B. Fadhil, A. W. Nayyef, and N. M. T. Al-Layla, "Biodiesel production from nonedible feedstock, radish seed oil by cosolvent method at room temperature: Evaluation and analysis of biodiesel," *Energy Sources*, Part A, vol. 42, pp. 1891–1901, 2020. https://doi.org/10.1080/15567036.2019.1604902

[11] M. M. Hassan and A. B. Fadhil, "Development of an effective solid base catalyst from potassium-based chicken bone (K-CBs) composite for biodiesel production from a mixture of non-edible feedstocks," *Energy Sources*, Part A, pp. 1–16, Informa UK Limited, 2021. https://doi.org/10.1080/15567036.2021.1927253

[12] J. Bhattacharjee and S. Roy, "Significance of renewable energy in water management and irrigation," in *Water Management in Developing Countries and Sustainable Development. Water Resources Development and Management*, Springer, 2024. https://doi.org/10.1007/978-981-99-8639-2_12

[13] Y. Guo, Z. Mustafaoglu, and D. Koundal, "Spam detection using bidirectional transformers and machine learning classifier algorithms," *Journal of Computational and Cognitive Engineering*, Apr. 22, 2022. https://doi.org/10.47852/bonviewJCCE2202192

[14] B. Najafi, S. Faizollahzadeh Ardabili, S. Shamshirband, K. Chau, and T. Rabczuk, "Application of ANNs, ANFIS and RSM to estimating and optimizing the parameters that affect the yield and cost of biodiesel production," *Engineering Applications of Computational Fluid Mechanics*, vol. 12, no. 1, pp. 611–624, Informa UK Limited, 2018. https://doi.org/10.1080/19942060.2018.1502688

[15] M. Abdelateef Mostafa, E. A. El-Hay, and M. M. ELkholy, "Recent trends in wind energy conversion system with grid integration based on soft computing methods: Comprehensive review, comparisons and insights," *Archives of Computational Methods in Engineering*, vol. 30, no. 3, pp. 1439–1478, Springer Science and Business Media LLC, 2022. https://doi.org/10.1007/s11831-022-09842-4

[16] M. Nazari-Heris, B. Mohammadi-Ivatloo, S. Asadi, J.-H. Kim, and Z. W. Geem, "Harmony search algorithm for energy system applications: An updated review and analysis," *Journal of Experimental & Theoretical Artificial Intelligence*, vol. 31, no. 5, pp. 723–749, Informa UK Limited, 2018. https://doi.org/10.1080/0952813x.2018.1550814

[17] M. Seraj et al., "Analytical research of artificial intelligent models for machining industry under varying environmental strategies: An industry 4.0 approach," *Sustainable Operations and Computers*, vol. 3, pp. 176–187, Elsevier BV, 2022. https://doi.org/10.1016/j.susoc.2022.01.006

[18] M. Fatima, N. U. K. Sherwani, S. Khan, and M. Z. Khan, "Assessing and predicting operation variables for doctors employing industry 4.0 in health care industry using an adaptive neuro-fuzzy inference system (ANFIS) approach," *Sustainable Operations and Computers*, vol. 3, pp. 286–295, Elsevier BV, 2022. https://doi.org/10.1016/j.susoc.2022.05.005

[19] A. J. Callejón-Ferre, B. Velázquez-Martí, J. A. López-Martínez, and F. Manzano-Agugliaro, "Greenhouse crop residues: Energy potential and models for the prediction of their higher heating value," *Renewable and Sustainable Energy Reviews*, vol. 15, no. 2, pp. 948–955, Elsevier BV, 2011. https://doi.org/10.1016/j.rser.2010.11.012

[20] A. K. Gupta, P. Kumar, R. K. Sahoo, A. K. Sahu, and S. K. Sarangi, "Performance measurement of plate fin heat exchanger by exploration: ANN, ANFIS, GA, and SA," *Journal of Computational Design and Engineering*, vol. 4, no. 1, pp. 60–68, Oxford University Press (OUP), 2016. https://doi.org/10.1016/j.jcde.2016.07.002
[21] A. Singh, M. Z. Khan, Yogesh, and P. Mahto, "The impact of low Reynolds number on coefficient of probe at different-different angle of S-type pitot tube," *Materials Today: Proceedings*, vol. 46, pp. 6867–6870, Elsevier BV, 2021. https://doi.org/10.1016/j.matpr.2021.04.443
[22] M. Aghbashlo, S. Hosseinpour, M. Tabatabaei, and M. Mojarab Soufiyan, "Multi-objective exergetic and technical optimization of a piezoelectric ultrasonic reactor applied to synthesize biodiesel from waste cooking oil (WCO) using soft computing techniques," *Fuel*, vol. 235, pp. 100–112, Elsevier BV, 2019. https://doi.org/10.1016/j.fuel.2018.07.095
[23] O. Khan, A. K. Yadav, M. E. Khan, and M. Parvez, "Characterization of bioethanol obtained from Eichhornia Crassipes plant; its emission and performance analysis on CI engine," *Energy Sources*, Part A, vol. 43, no. 14, pp. 1793–1803, Informa UK Limited, 2019. https://doi.org/10.1080/15567036.2019.1648600
[24] A. K. Yadav, O. Khan, and M. E. Khan, "Utilization of high FFA landfill waste (leachates) as a feedstock for sustainable biodiesel production: Its characterization and engine performance evaluation," *Environmental Science and Pollution Research*, vol. 25, no. 32, pp. 32312–32320, Springer Science and Business Media LLC, 2018. https://doi.org/10.1007/s11356-018-3199-0
[25] C. Li, J. M. Mogollón, A. Tukker, and B. Steubing, "Environmental impacts of global offshore wind energy development until 2040," *Environmental Science & Technology*, vol. 56, no. 16, pp. 11567–11577, American Chemical Society (ACS), 2022 https://doi.org/10.1021/acs.est.2c02183
[26] Z. Cao et al., "Resourcing the fairytale country with wind power: A dynamic material flow analysis," *Environmental Science & Technology*, vol. 53, no. 19, pp. 11313–11322, American Chemical Society (ACS), 2019. https://doi.org/10.1021/acs.est.9b03765
[27] C. K. Lee, H. H. Khoo, and R. B. H. Tan, "Life cyle assessment based environmental performance comparison of batch and continuous processing: A case of 4-d-erythronolactone synthesis," *Organic Process Research & Development*, vol. 20, no. 11, pp. 1937–1948, American Chemical Society (ACS), 2016. https://doi.org/10.1021/acs.oprd.6b00275
[28] E. E. Kwon, S.-H. Cho, and S. Kim, "Synergetic sustainability enhancement via utilization of carbon dioxide as carbon neutral chemical feedstock in the thermo-chemical processing of biomass," *Environmental Science & Technology*, vol. 49, no. 8, pp. 5028–5034, American Chemical Society (ACS), 2015. https://doi.org/10.1021/es505744n
[29] B. A. Wender et al., "Illustrating anticipatory life cycle assessment for emerging photovoltaic technologies," *Environmental Science & Technology*, vol. 48, no. 18, pp. 10531–10538, 2014. https://doi.org/10.1021/es5016923
[30] A. A. Chhipa et al., "Adaptive neuro-fuzzy inference system-based maximum power tracking controller for variable speed WECS," *Energies*, vol. 14, no. 19, p. 6275, MDPI AG, 2021. https://doi.org/10.3390/en14196275
[31] Q. Sun, G. Li, L. Duan, and Z. He, "The coupling of tower-shadow effect and surge motion intensifies aerodynamic load variability in downwind floating

offshore wind turbines," *Energy*, vol. 282, p. 128788, Elsevier BV, 2023. https://doi.org/10.1016/j.energy.2023.128788
[32] R. Collobert et al., "Natural language processing (almost) from scratch," *Journal of Machine Learning Research*, vol. 12, pp. 2493–2537, 2011. [Online]. Available: https://arxiv.org/abs/1103.0398
[33] N. Yang et al., "Decarbonization of the wind power sector in China: Evolving trend and driving factors," *Environmental Impact Assessment Review*, vol. 103, p. 107292, Elsevier BV, 2023. https://doi.org/10.1016/j.eiar.2023.107292
[34] S. Langkau et al., "A stepwise approach for Scenario-based Inventory Modelling for Prospective LCA (SIMPL)," *The International Journal of Life Cycle Assessment*, vol. 28, no. 9, pp. 1169–1193, Springer Science and Business Media LLC, 2023. https://doi.org/10.1007/s11367-023-02175-9
[35] D. Clarabut et al., "The nonlinear effects of spinning on the dynamics of a pitching cantilever," *Journal of Sound and Vibration*, vol. 569, p. 117876, 2024. https://doi.org/10.1016/j.jsv.2023.117876
[36] I. Shafi et al., "An artificial neural network-based approach for real-time hybrid wind – solar resource assessment and power estimation," *Energies*, vol. 16, no. 10, p. 4171, MDPI AG, 2023. https://doi.org/10.3390/en16104171
[37] M. Yildiz, U. Kale, and A. Nagy, "Analyzing well-to-pump emissions of electric and conventional jet fuel for aircraft propulsion," *Aircraft Engineering and Aerospace Technology*, vol. 94, no. 10, pp. 1605–1613, Emerald, 2022. https://doi.org/10.1108/aeat-02-2021-0032
[38] J. K. Nøland et al., "Spatial energy density of large-scale electricity generation from power sources worldwide," *Scientific Reports*, vol. 12, no. 1, Springer Science and Business Media LLC, 2022. https://doi.org/10.1038/s41598-022-25341-9
[39] F. Abderrahmane et al., "An improved integrated maintenance/spare parts management for wind turbine systems with adopting switching concept," *Energy Reports*, vol. 8, pp. 936–955, Elsevier BV, 2022. https://doi.org/10.1016/j.egyr.2022.07.123
[40] A. Guediri, M. Hettiri, and A. Guediri, "Modeling of a wind power system using the genetic algorithm based on a doubly fed induction generator for the supply of power to the electrical grid," *Processes*, vol. 11, no. 3, p. 952, MDPI AG, 2023. https://doi.org/10.3390/pr11030952
[41] R. L. Vekariya et al., "An overview of engineered porous material for energy applications: A mini-review," *Ionics*, vol. 24, pp. 1–17, 2018. https://doi.org/10.1007/s11581-017-2338-9
[42] K. Xu et al., "A comprehensive estimate of life cycle greenhouse gas emissions from onshore wind energy in China," *Journal of Cleaner Production*, vol. 338, p. 130683, Elsevier BV, 2022. https://doi.org/10.1016/j.jclepro.2022.130683
[43] K. Aganovic and S. Smetana, "Environmental impact assessment of pulsed electric fields technology for food processing," in *Pulsed Electric Fields Technology for the Food Industry*, pp. 521–539, Springer International Publishing, 2022. https://doi.org/10.1007/978-3-030-70586-2_19
[44] O. Adeyomoye, C. Akintayo, K. Omotuyi, and A. Adewumi, "The biological roles of urea: A review of preclinical studies," *Indian Journal of Nephrology*, vol. 32, no. 6, p. 539, Medknow, 2022. https://doi.org/10.4103/ijn.ijn_88_21

[45] H. Bahlawan, M. Morini, M. Pinelli, P. R. Spina, and M. Venturini, "Simultaneous optimization of the design and operation of multi-generation energy systems based on life cycle energy and economic assessment," *Energy Conversion and Management*, vol. 249, p. 114883, Elsevier BV, 2021. https://doi.org/10.1016/j.enconman.2021.114883
[46] H. Du et al., "Modeling of cyber attacks against converter-driven stability of PMSG-based wind farms with intentional subsynchronous resonance," in *2021 IEEE International Conference on Communications, Control, and Computing Technologies for Smart Grids (SmartGridComm)*, IEEE, 2021. https://doi.org/10.1109/smartgridcomm51999.2021.9632318
[47] M. Pizzol, R. Sacchi, S. Köhler, and A. Anderson Erjavec, "Non-linearity in the Life cycle assessment of scalable and emerging technologies," *Frontiers in Sustainability*, vol. 1, Frontiers Media SA, 2021. https://doi.org/10.3389/frsus.2020.611593
[48] M. K. van der Hulst et al., "A systematic approach to assess the environmental impact of emerging technologies: A case study for the GHG footprint of CIGS solar photovoltaic laminate," *Journal of Industrial Ecology*, vol. 24, no. 6, pp. 1234–1249, Wiley, 2020. https://doi.org/10.1111/jiec.13027
[49] G. Santo, M. Peeters, W. Van Paepegem, and J. Degroote, "Effect of rotor – tower interaction, tilt angle, and yaw misalignment on the aeroelasticity of a large horizontal axis wind turbine with composite blades," *Wind Energy*, vol. 23, no. 7, pp. 1578–1595, Wiley, 2020. https://doi.org/10.1002/we.2501
[50] A. Awada, R. Younes, and A. Ilinca, "Review of vibration control methods for wind turbines," *Energies*, vol. 14, no. 11, p. 3058, MDPI AG, 2021. https://doi.org/10.3390/en14113058
[51] S. Verma, A. R. Paul, and N. Haque, "Selected environmental impact indicators assessment of wind energy in india using a life cycle assessment," *Energies*, vol. 15, no. 11, p. 3944, MDPI AG, 2022. https://doi.org/10.3390/en15113944
[52] D. M. Ngoc, M. Luengchavanon, P. T. Anh, K. Humphreys, and K. Techato, "Shades of green: Life cycle assessment of a novel small-scale vertical axis wind turbine tree," *Energies*, vol. 15, no. 20, p. 7530, MDPI AG, 2022. https://doi.org/10.3390/en15207530
[53] V. Buhrmester, D. Münch, and M. Arens, "Analysis of explainers of black box deep neural networks for computer vision: A survey," *Machine Learning and Knowledge Extraction*, vol. 3, no. 4, pp. 966–989, MDPI AG, 2021. https://doi.org/10.3390/make3040048
[54] J. Bhattacharjee and S. Roy, "Utilizing a variable material approach to combat climate change," *Material Science Research India*, vol. 20, no. 3, pp. 141–145, Oriental Scientific Publishing Company, 2024. https://doi.org/10.13005/msri/200301
[55] R. Garg and A. Anjum, Eds., "Smart and sustainable applications of nanocomposites," *Advances in Chemical and Materials Engineering*, IGI Global, 2024. https://doi.org/10.4018/979-8-3693-1094-6
[56] J. Bhattacharjee and S. Roy, "Review on green resources and AI for biogenic solar power," *Energy Storage and Conversion*, vol. 2, 2024. https://ojs.acadpub.com/index.php/ESC/article/view/457
[57] A. De, J. Bhattacharjee, S. R. Chowdhury, and S. Roy, "A comprehensive review on third-generation photovoltaic technologies," *Journal of Chemical Engineering Research Updates*, vol. 10, pp. 1–17, Avanti Publishers, 2023. https://doi.org/10.15377/2409-983x.2023.10.1

[58] J. Bhattacharjee, and S. Roy, "Biogenic electrodes for low-temperature solid oxide fuel cells," *ECS Meeting Abstracts*, vol. MA2023–01, no. 40, pp. 2805–2805, The Electrochemical Society, 2023. https://doi.org/10.1149/ma2023-01402805mtgabs
[59] J. Bhattacharjee and S. Roy, "Green remediation of microplastics using biona-nomaterials," in *Remediation of Plastic and Microplastic Waste*, pp. 240–260, CRC Press, 2023 https://doi.org/10.1201/9781003449133-14
[60] S. Roy, R. Chatterjee, and S. B. Majumder, "Magnetoelectric coupling in sol-gel synthesized dilute magnetostrictive-piezoelectric composite thin films," *Journal of Applied Physics*, vol. 110, no. 3, AIP Publishing, 2011. https://doi.org/10.1063/1.3610795
[61] S. S. Band, S. Ardabili, A. Mosavi, C. Jun, H. Khoshkam, and M. Moslehpour, "Feasibility of soft computing techniques for estimating the long-term mean monthly wind speed," *Energy Reports*, vol. 8. Elsevier BV, pp. 638–648, Nov. 2022. https://doi.org/10.1016/j.egyr.2021.11.247

Chapter 7

Artificial Neural Network-Based MPPT Controller for Variable-Speed Wind Energy Conversion System

Mahdi Hermassi, Saber Krim, Youssef Kraiem, and Mohamed Ali Hajjaji

7.1 INTRODUCTION

Recently, there has been a growing interest in wind generation systems as environmentally friendly and secure sources of renewable power [1, 2]. Wind energy conversion systems (WECSs) can operate, using power electronic converters, in either fixed or variable speed (VS) modes. VS WECSs offer several advantages over fixed-speed generation. These advantages include operating at the maximum power point (MPP), achieving enhanced effectiveness, increasing energy capacity, and improving power quality [3]. Interestingly, among the diverse range of VS wind turbines (WTs), permanent magnet synchronous generator (PMSG) based WECSs have been extensively applied in wind power production thanks to their compact design, substantial power output in relation to size, and impressive torque-inertia relationship [3]. This progress has sparked a significant increase in the research interest centered around PMSG-based WECSs for wind energy generation [4]. Nonetheless, as a result of the unpredictable and fluctuating nature of wind speed (WS), the power output of WTs can still experience variations. A maximum power point tracking (MPPT) controller assumes a pivotal role in optimizing the generator speed and output power measurements of a wind energy system. This system is acknowledged for its significant nonlinearity and distinct, sudden shifts in WS. The primary goal is to optimize the power output of the system, regardless of fluctuations in WS.

The operational range of a VS WECS can be divided into four zones, as depicted in Figure 7.1. Zone I signifies low WS conditions that lack the capacity for generating electric power. In Zone II, the system is optimized to maximize power production by dynamically adjusting to the WS through the implementation of the MPPT algorithm, all while maintaining the optimal blade pitch angle (β). In Zone III, when the WS exceeds its designated value, adjustments to the pitch angle β are made to maintain the generated power around the nominal level. Zone IV corresponds to high WS that has the potential to cause damage to the WT [5]. Consequently, protective measures are implemented, such as emergency devices that halt the turbine operation.

DOI: 10.1201/9781003462460-7

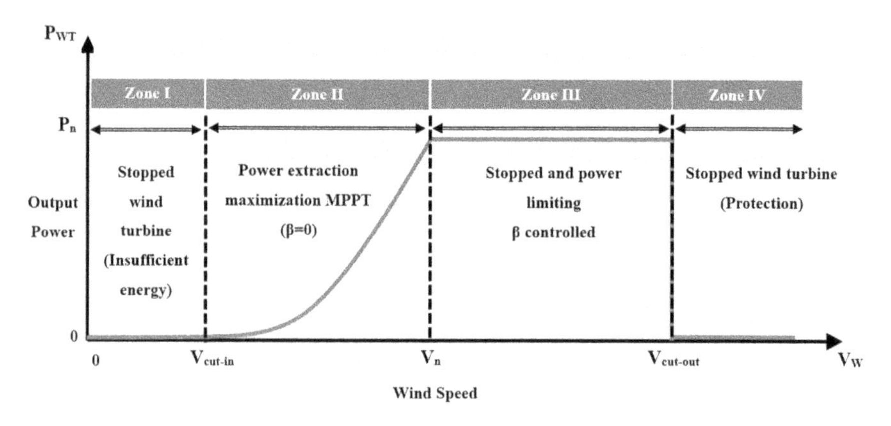

Figure 7.1 Various operational zones of WT.

This chapter focuses on Zone II, specifically investigating the utilization of an artificial neural network-based MPPT (ANN-MPPT) algorithm.

Various MPPT algorithms have been extensively discussed in existing literature [6–8]. These algorithms can be divided into two main groups: direct power control-based MPPT algorithms, which comprise the optimum relation-based (ORB), hill climbing search (HCS), and incremental conductance (INC), and indirect power control-based MPPT algorithms, which include power signal feedback (PSF), optimal torque control (OTC), and tip-speed ratio (TSR) methods.

The OTC method involves modifying the generator torque using the primary power reference torque associated with the current WS conditions [9]. This approach is distinguished by its quick responsiveness, efficiency, and straightforward nature. Due to the lack of direct WS measurements, alterations in WS are not mirrored in the reference signal [7].

The INC methods function independently of the sensor prerequisites and the distinct attributes of WTs and generators. As a result, systems utilizing these methods experience reduced costs and improved reliability [10]. As indicated by the authors in the references [11, 12], the determination of the MPPT operating point can rely on the power-speed slope. However, this technique becomes unstable when the WT inertia changes within a variable speed (VS) wind scenario, as outlined in reference [13]. To tackle this instability problem at various WS, a novel approach called fractional order INC was introduced in [12]. This method incorporated a variable step size to efficiently track the maximum power point under changing wind conditions.

The MPPT method based on the PSF employs a power control loop that incorporates data from the WTs maximum power curve [14]. On the other hand, the TSR-based MPPT approach is straightforward to construct and boasts high efficiency. The limitation of this algorithm lies in its requirement for an optimal power coefficient and an optimal TSR [15].

The HCS method is characterized by both its unpredictability and robustness, hinging on prior knowledge of WT attributes. This method furnishes the local maximum point for a specified function. However, a notable limitation of this approach arises from its susceptibility to incorrectly determining the optimal path in the event of a sudden shift in wind direction [16].

The ORB-based MPPT approach relies on a precise correlation between speed, WT power output, converter DC voltage, current, and power. Its key advantages are the absence of a sensor requirement and the ability to perform fast tracking [17]. However, to implement this technique, it is necessary to have characteristic curves that provide insights into turbine power and converter DC current at various WS. Consequently, the observation of the optimal current curve becomes crucial for effectively tracking the MPP [18].

These approaches are designed to optimize turbine operation and track the MPP, ensuring efficient energy conversion from the wind source. However, a limitation of these approaches is their restricted capacity to manage significant power fluctuations resulting from changes in wind conditions, potentially leading to misinterpretations by the MPPT strategy.

Computational intelligence techniques commonly employ soft computing approaches, such as fuzzy logic and neural networks, to address complex and uncertain problems that are difficult to solve using traditional rule-based programming or mathematical methods [19]. These methods enable machines to learn from data, make decisions, and optimize solutions in dynamic and uncertain environments, much like how human intelligence operates. These advanced controllers demonstrate swift responses even when dealing with abrupt changes in WS [20]. Furthermore, these soft computing-based MPPT algorithms consistently maintain high accuracy across various WS conditions. However, the fuzzy logic controller-based MPPT algorithm requires prior system knowledge, and its control algorithm is complex. Conversely, improved controllers like the artificial neural network (ANN) have gained popularity for managing nonlinear systems, offering enhanced stability. These controllers ensure fast convergence and feature a simple network structure [21–24]. The ANN introduces an alternative approach to identify the peak of maximum power by processing diverse input variables [25]. Each neural network is comprised of an input layer, a hidden layer, and an output layer, with the flexibility to adjust the number of nodes as required. Utilizing the ANN-based controller provides a more effective and dependable approach, surpassing traditional controllers in extracting the maximum power from the kinetic energy of the wind.

The main objective of this chapter is to present an MPPT system with speed control for WTs. This study proposes three different MPPT algorithms to track a set point that is dependent on the WS. These algorithms include PI linear control, nonlinear sliding mode control (SMC), and the proposed ANN controllers. The examination of the results indicates that the ANN developed in this study surpasses the performance of other controllers in terms of precision, reference tracking, and efficiency enhancement. The ANN demonstrates high accuracy and significantly improves the overall performance of the MPPT mechanism.

The structure of this chapter is outlined as follows: Following the introduction, the Section 7.2 delves into the modeling of the WT and its mechanical components. The subsequent Section 7.3 expounds on the suggested control approaches. Section 7.4 provides an in-depth exploration of the obtained outcomes. Section 7.5 summarizes the findings.

7.2 WIND GENERATOR MODELING

The expression for the mechanical power P_t available on the shaft of a WT is as follows [26]:

$$P_t = \frac{1}{2}\rho\pi R_t^2 V^3 C_p(\lambda, \beta) \tag{7.1}$$

The relative speed, denoted as λ, can be defined as the ratio between the WS and the speed of the WT:

$$\lambda = \frac{R_t \Omega_t}{V} \tag{7.2}$$

where ρ(kg/m^3) represents the air density, C_p denotes the power coefficient, β represents the blade pitch angle, Ω_g (rad/s) denotes the rotational speed of the turbine shaft, R_t represents the turbine radius, and V denotes the WS.

The power coefficient C_p serves as an indicator of the aerodynamic effectiveness of the WT and is influenced by its characteristics. In our study, we utilize the following formula for the power coefficient that is functional with the relative speed and the pitch angle [27]:

$$C_p = 0.5\left(\frac{151}{\lambda_i} - 0.58\beta - 0.002\beta^{2.14} - 10\right)e^{-\frac{18.4}{\lambda_i}} \tag{7.3}$$

$$\lambda_i = \frac{1}{\frac{1}{\lambda - 0.02\beta} - \frac{0.003}{\beta^3 + 1}} \tag{7.4}$$

The relationship between the power coefficient C_p and the λ curve is illustrated in Figure 7.2, using Equations (7.3) and (7.4). The power coefficient C_p exhibits an optimal value that leads to maximum power extraction C_{pmax}. By adjusting λ to its optimal value λ_{opt}, the maximum power coefficient can be attained when β=0. From Equations (7.1) and (7.2), the following expression is obtained:

$$P_{t\max} = \frac{1}{2}\rho\pi R_t^5\left(\frac{C_{p\max}}{\lambda^3_{opt}}\right)\Omega^3_{opt} \tag{7.5}$$

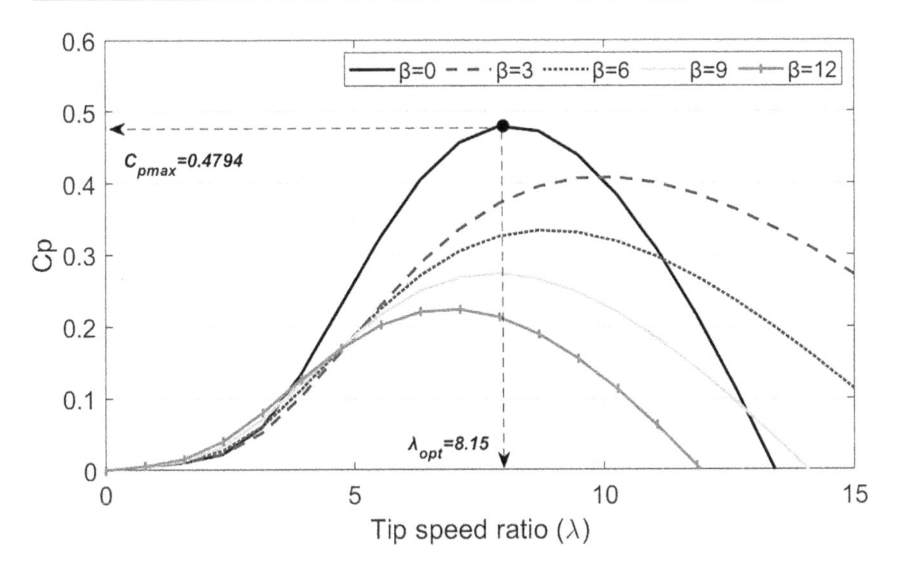

Figure 7.2 C_p versus λ.

$$\Omega_{opt} = \frac{\lambda_{opt} V}{R_t} \tag{7.6}$$

In this study, the WT under examination is linked to a PMSG to transform mechanical energy into electrical energy. The diagram presented in Figure 7.3 summarizes the model of the WT.

The PMSG model can be expressed using the following equations, based on the d-q coordinate system in the Park synchronized rotation frame [28, 29].

$$\begin{cases} \dfrac{d}{dt} i_{sd} = \dfrac{1}{L_s}\left(V_{sd} - R_s i_{sd} + \omega L_s i_{sq}\right) \\ \dfrac{d}{dt} i_{sq} = \dfrac{1}{L_s}\left(V_{sq} - R_s i_{sq} - \omega L_s i_{sd} - \omega \psi_m\right) \end{cases} \tag{7.7}$$

where R_s and L_s are respectively the stator resistance and the stator inductance, i_{sq} and i_{sd} are respectively the quadrature components and direct of the stator current, V_{sq} and V_{sd} are respectively the quadrature components and direct of the stator voltage, Ψ_m represents the flux of the permanent magnet, and p denotes the number of pole pairs.

To express the PMSG in the d-q frame, the electromagnetic torque T_{em} is formulated as follows:

$$T_{em} = p \Psi_m i_{sq} \tag{7.8}$$

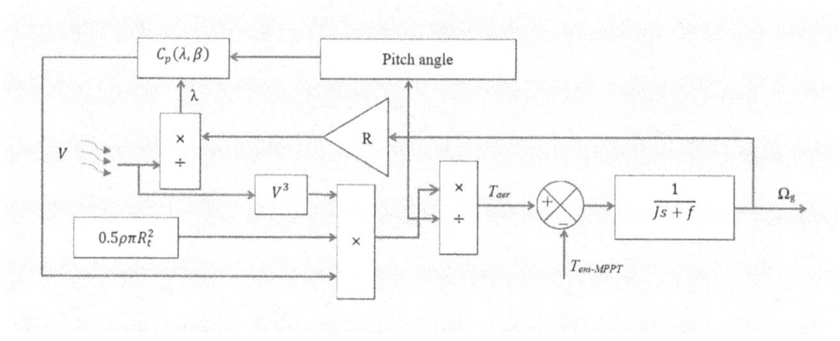

Figure 7.3 Diagram of WT model.

The following equation presents the mechanical rotational speed of the generator:

$$j\frac{d}{dt}\Omega_g = T_{aer} - T_{em} - f\Omega_g \tag{7.9}$$

where j and f are the inertia of the shaft and the coefficient of friction, respectively.

7.3 MPPT WITH SPEED CONTROL

In this part, our objective is to adjust the generator's electromagnetic torque to regulate the mechanical speed towards a reference value. For this purpose, a speed control mechanism for the generator is necessary.

Achieving maximal mechanical power involves operating the system at its associated peak power coefficient value C_p and optimal TSR λ_{opt}. As a result, the desired speed of the generator (Ω^*_g) is expressed using the following equation:

$$\Omega^*_g = \frac{\lambda_{opt} V}{R_t} G \tag{7.10}$$

The structure of the suggested MPPT algorithm is depicted in Figure 7.4. In this scheme, the electromagnetic torque reference $T_{em\text{-}MPPT}$ responsible for controlling the speed of the PMSG, is acquired from the output of the speed controller.

7.3.1 PI Controller-Based MPPT

In the assumption of an ideal machine and drive system, regardless of the desired power to be generated, the electromagnetic torque is set equal to its

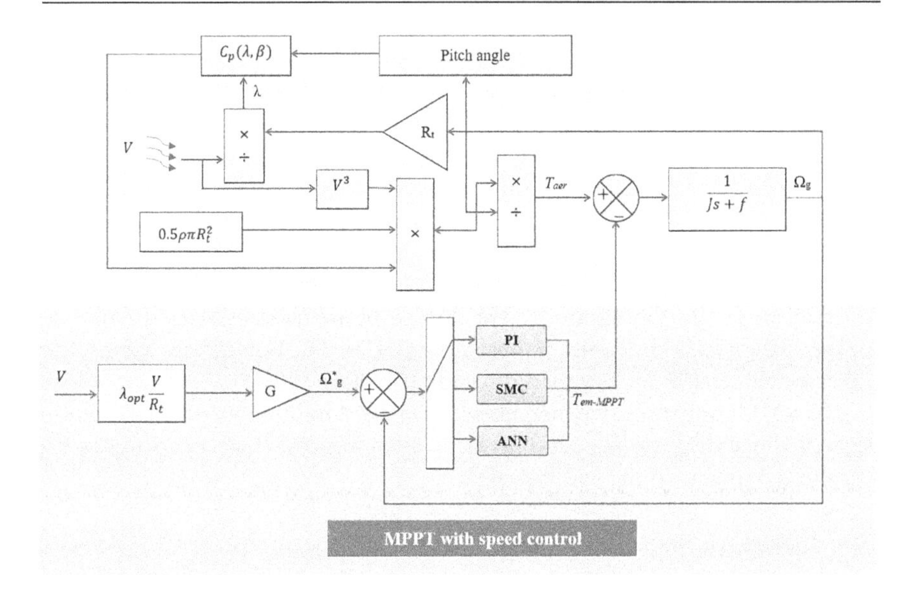

Figure 7.4 Diagram of MPPT algorithm with speed control.

reference value. This implies that the system operates with perfect efficiency and that the torque applied to the generator precisely matches the desired value.

$$T_{em} = T_{em-MPPT} \tag{7.11}$$

The reference electromagnetic torque is determined as follow:

$$T_{em-MPPT} = PI.\left(\Omega^*_g - \Omega_g\right) \tag{7.12}$$

$$\Omega^*_g = \Omega_{Turbine-ref}.G \tag{7.13}$$

$$\Omega_{Turbine-ref} = \lambda_{opt}\frac{V}{R_t} \tag{7.14}$$

with

$$PI = K_p e(t) + K_i \int e(t)dt \tag{7.15}$$

where $e(t) = \Omega^*_g(t) - \Omega_g(t)$.

Additionally, during digital simulation, the parameters of the PI controller are chosen through practical simulations. This iterative process involves adjusting the K_p and K_i gains to optimize the controller's performance.

7.3.2 SMC-Based MPPT

The SMC algorithm combines nonlinear control with variable structure methods. It utilizes various control structures designed to ensure that system trajectories consistently converge to a specified adjacent region of the sliding structure [27]. This region represents the system's normal behavior and stability. The SMC strategy aims to maintain the system within the desired operating region despite disturbances or uncertainties.

As the system traverses the limits of the control structures, it enters a state referred to as the sliding mode. The sliding mode characterizes the motion of the system as it glides along these boundaries. The boundaries themselves form a geometrical locus called the sliding surface [30].

The SMC comprises two elements: a non-continuous control based on the sign of the sliding surface, and an equivalent control that encapsulates the system dynamics on the sliding surface.

$$T_{em} = T_{em-eq} + T_{em-n} \tag{7.16}$$

To design the command $T_{em\text{-}ref}$, the surface's relative degree is set to 1. The sliding surface is determined as follows:

$$S(\Omega_g) = \Omega_g^* - \Omega_g \tag{7.17}$$

The $T_{em\text{-}eq}$ command's equivalent portion represents an ideal sliding motion, disregarding any disturbances and uncertainties in the system. In physical terms, it corresponds to the mean value of the actual command and is derived from the invariance conditions of the sliding surface.

$$S(\Omega_g) = \frac{d}{dt} S(\Omega_g) = 0 \tag{7.18}$$

From Equation (7.17), we can derive the surface's derivative as follows:

$$\dot{S}(\Omega_g) = \dot{\Omega}_g^* - \dot{\Omega}_g \tag{7.19}$$

Substituting Equation (7.9) into Equation (7.19), the resulting equation is:

$$\dot{S}(\Omega_g) = \dot{\Omega}_g^* + \frac{1}{j}\left(T_{em} + f_v \Omega_g - T_g\right) \tag{7.20}$$

By substituting the equation of T_{em} with the sum of equivalent commands $T_{emeq} + T_{emn}$ in Equation (7.20), the resulting equation can be derived:

$$\dot{S}(\Omega_g) = \Omega_g^* + \frac{1}{j}\left(\left(T_{emeq} + T_{emn}\right) + f_v \Omega_g - T_g\right) \tag{7.21}$$

Under steady state conditions and during the sliding mode, the following relationships hold: $T_{emn} = 0$, $S(\Omega g) = 0$ and $\dot{S}(\Omega_g) = 0$. By examining the earlier equations, we can extract the formulation for the equivalent command, T_{emeq}:

$$T_{emeq} = -j.\dot{\Omega}_g^* - f_v.\Omega_g + T_g \tag{7.22}$$

To establish stability and analyze the system's behavior, the Lyapunov function is chosen as follows [27]:

$$V\left(S\left(\Omega_g\right)\right) = \frac{1}{2}S\left(\Omega_g\right)^2 \tag{7.23}$$

The time derivative of the chosen Lyapunov function is expressed as follows:

$$\dot{V}\left(S\left(\Omega_g\right)\right) = S\left(\Omega_g\right).\dot{S}\left(\Omega_g\right) \prec 0 \tag{7.24}$$

with $\dot{S}\left(\Omega_g\right) = \dot{\Omega}_g^* - \dot{\Omega}_g$

To satisfy the attractiveness condition as specified in Equation (7.24), it is sufficient to choose the discontinuous component of the order as follows:

$$T_{em-n} = -K_g sign\left(S\left(\Omega_g\right)\right) \tag{7.25}$$

where $K_g>0$ represents the command gain, and sign $(S(\Omega_g))$ is the sign function defined as follows:

$$sign\left(S\left(\Omega_g\right)\right)\begin{cases} 1 \quad if \quad S(x) \succ \varepsilon \\ S\left(\Omega_g\right) \quad if \quad -\varepsilon \prec \quad S\left(\Omega_g\right) \prec \varepsilon \\ -1 \quad if \quad S\left(\Omega_g\right) \prec -\varepsilon \end{cases} \tag{7.26}$$

By substituting Equations (7.21) into Equation (7.22), the derivative of the sliding surface can be reformulated as:

$$\dot{S}\left(\Omega_g\right) = \frac{1}{j}T_{emn} \tag{7.27}$$

Finally, Equation (7.24) becomes:

$$\begin{aligned} \dot{V}\left(S\left(\Omega_g\right)\right) &= S\left(\Omega_g\right).\dot{S}\left(\Omega_g\right) = S\left(\Omega_g\right)\dot{\Omega}_g^* + \frac{1}{j}\left(\left(T_{emeq} + T_{emn}\right) + f_v\Omega_g - T_g\right) \\ &= S\left(\Omega_g\right)\dot{\Omega}_g^* + \frac{1}{j}\left(-j.\dot{\Omega}_g^* - f_v.\Omega_g + T_g - K_g sign\left(S\left(\Omega_g\right)\right) + f_v\Omega_g - T_g\right) \\ &= -K_g\frac{1}{j}S\left(\Omega_g\right)sign\left(S\left(\Omega_g\right)\right) \prec 0 \end{aligned} \tag{7.28}$$

Hence, the system stability is achieved.

7.3.3 Design of Suggested ANN-Based MPPT Strategy

Primarily, an ANN is a machine learning model that takes inspiration from the structure and functionality of the human brain. This algorithm consists of interconnected layers of neurons that undertake the task of information processing and transmission [31]. The process starts when the ANN receives an input. This processed input then goes through several concealed layers before getting an output [32]. In this arrangement, each hidden layer applies specific weights to the received input before relaying it to the subsequent layer. During the training phase, these weights are fine-tuned, allowing the network to acquire knowledge and enhance its performance over time. The fundamental concept of an ANN, rooted in the mathematical neuron, is depicted in Figure 7.5. An ANN is a dynamic system, often with nonlinear characteristics, which can execute inter input/output functions.

In this study, the ANN-based MPPT technique is proposed for regulating the mechanical speed of the WT. The objective is to ensure the maximum available power extraction amidst fluctuations in WS. The suggested ANN-based MPPT algorithm uses the measured rotor speed and its reference as input parameters. The ANN network then calculates and provides the instantaneous torque reference as the output.

An ANN is organized into three layers:

- The input layer is responsible for the initial distribution of the input vector to the hidden layer, without substantial processing.
- The hidden layer serves as the computational center of the ANN.

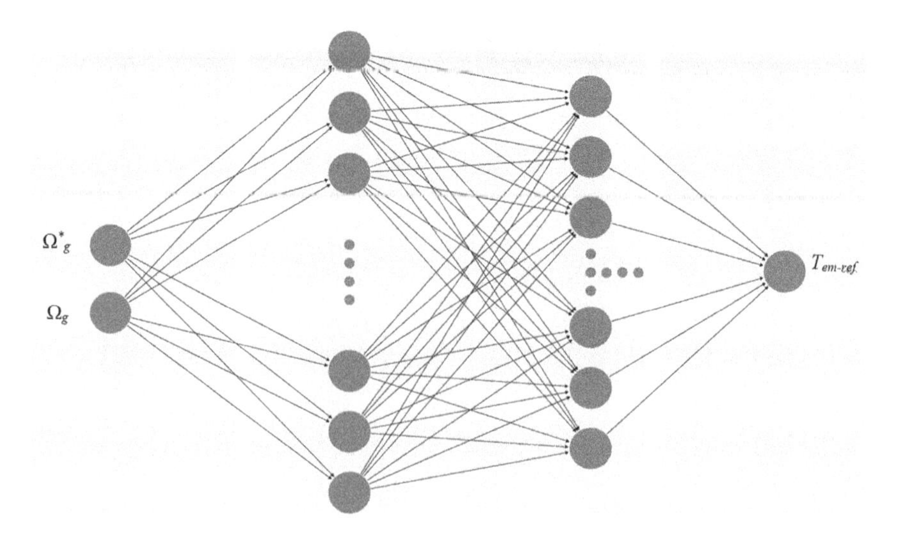

Figure 7.5 Architecture of proposed ANN.

- The output layer is where the cumulative outputs from the hidden layer are synthesized.

The initial layer consists of two neurons, the middle layer contains ten neurons, and the final layer has one neuron. To determine the optimal count of hidden layers and their respective neurons, an empirical approach is employed. This decision is based on achieving the required level of precision for the proposed methodology. Figure 7.6 illustrates the configuration of the neural network controller put forward in this study. The dataset is divided as follows: 70% of set points are allocated for training, while 15% for testing and 15 % for validation. Following this data allocation, a nonlinear network utilizing regression analysis is utilized to evaluate the effectiveness of the ANN controller, as depicted in Figure 7.8.

The training phase is a crucial step, indispensable for fine-tuning the parameters of the ANN structure to obtain the most accurate and reliable output. Through iterative adjustments to weights and biases, the network learns from the provided data, ensuring its adaptability to the intricate dynamics of the WECS. Moreover, we have incorporated an offline step into our training process, where different sets of neural network parameters, including the number of neurons, structural configuration, activation function, and training algorithm, are systematically explored.

The ANN training procedure specifically employs the Levenberg-Marquardt method, a resilient and rapidly converging algorithm, for optimizing the quadratic error. After multiple iterations, the results, as illustrated in Figures 7.7 and 7.8, underscore the method's capability to reach the desired error level with high approximation capacity and rapid convergence. It is imperative to highlight that our parameter selection is not purely theoretical but is rooted in practical considerations, with the training phase serving as a key mechanism to enhance the ANN's ability to generalize and produce precise predictions. This comprehensive approach ensures that the selected parameters align with the complexity of the WECS, contributing to the overall robustness and efficacy of our proposed methodology.

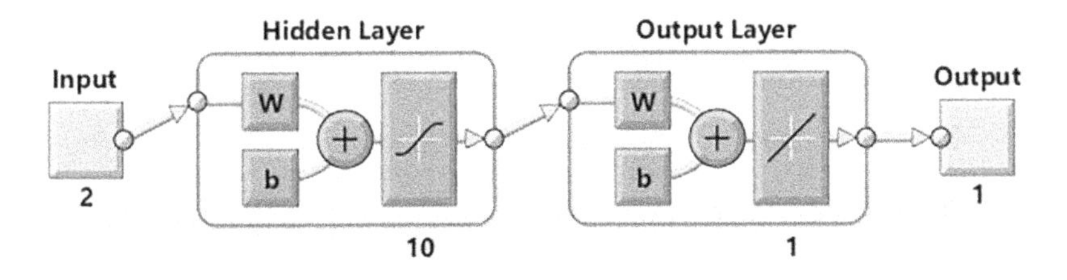

Figure 7.6 ANN configuration.

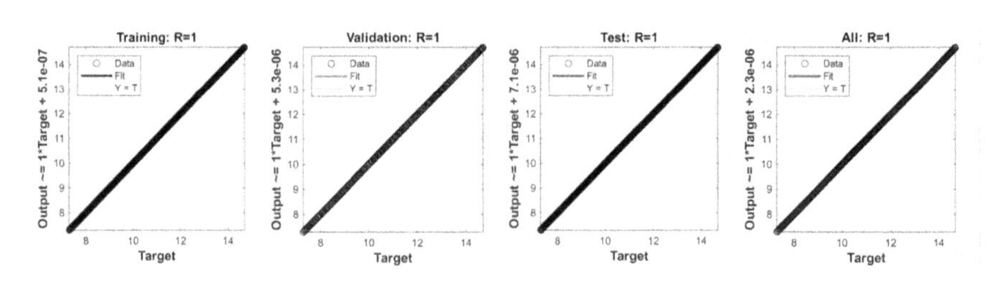

Figure 7.7 Target and output fitting correlations.

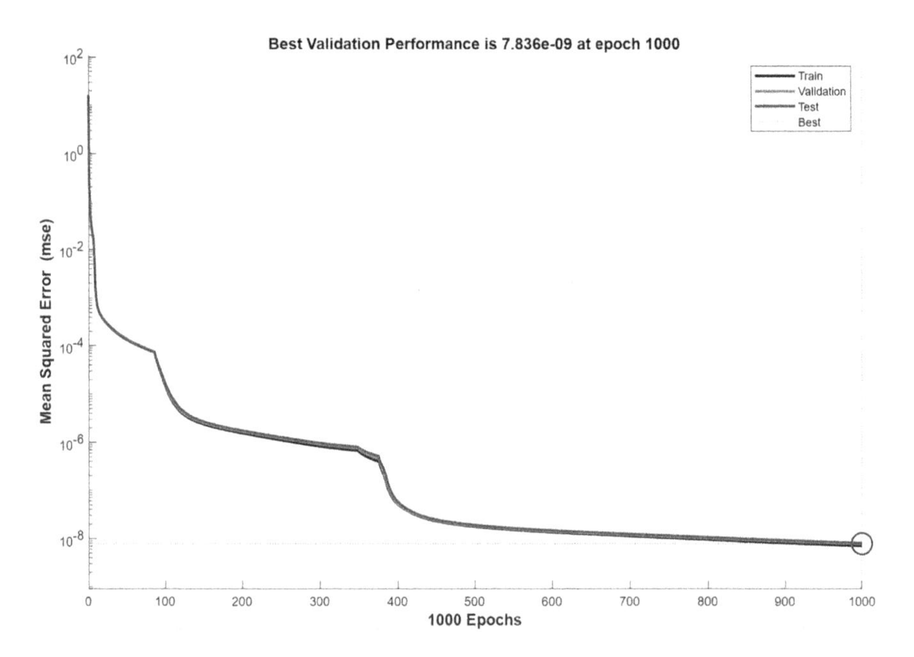

Figure 7.8 Training, validation, and testing results error for 1000 epochs.

7.4 SIMULATION RESULTS AND DISCUSSION

The performance of the selected controllers – PI, SMC, and the proposed ANN – is evaluated in the wind system under two wind profiles. The first profile involves a step change in wind speed (WS), and the second presents random variations in WS. The goal is to analyze the attributes of the suggested controllers, including time responses, tracking precision, and efficiency.

7.4.1 Case 1: MPPT Performance under Step Wind Profile Variation

The outcomes highlight the enhanced dynamic performance of the ANN controller-based MPPT compared to PI and SMC controllers, particularly in the presence of moderate WS disruptions, as depicted in Figure 7.9. In these diverse scenarios, it is worth noting that the power coefficient C_p (as depicted in Figure 7.10) achieves its maximum value of 0.4794 when the blade pitch angle is sustained at its minimal designated setting ($\beta = 0$) and the speed ratio is set to $\lambda_{opt} = 8.15$.

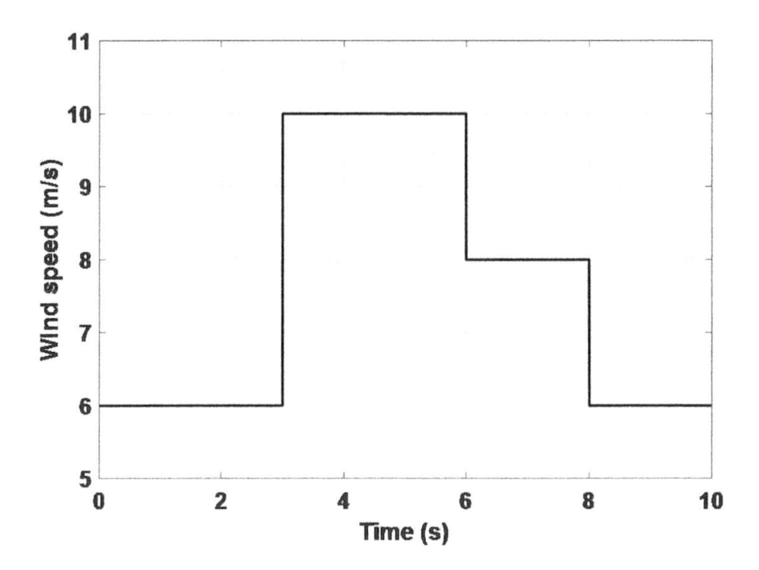

Figure 7.9 Step WS.

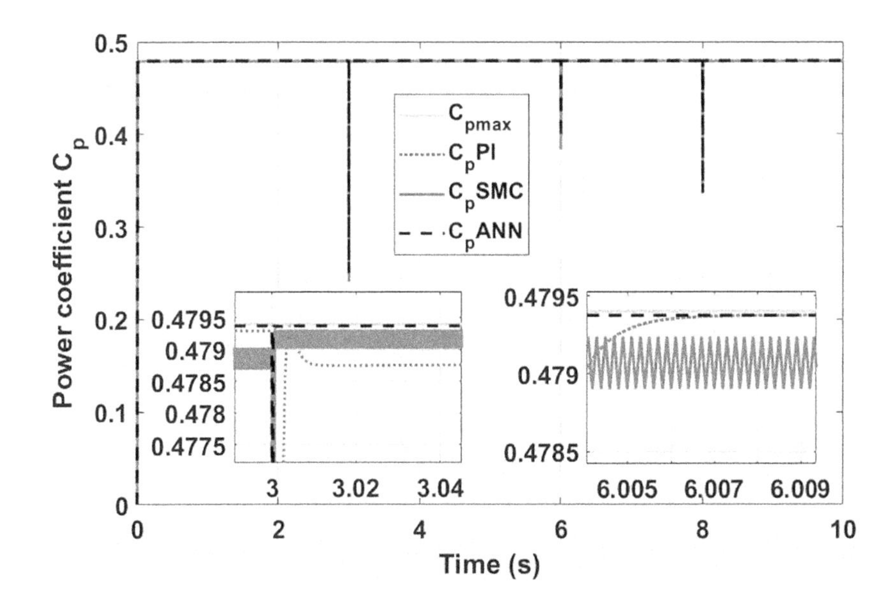

Figure 7.10 Power coefficient C_p for step WS.

The high error observed close to the starting time (time zero) is attributed to the significant damping present during the initial turbine operation.

The results of the MPPT simulation, integrating mechanical speed control with the three suggested controllers (PI, SMC, ANN), unequivocally demonstrate that, at every wind speed level, the mechanical speeds impeccably align with their designated references for all three MPPT algorithms, as depicted in Figure 7.11. Nevertheless, the PI controller is identified as the slowest and least efficient method and as poorly performing in terms of precision and accuracy tracking, indicating the limitations of the traditional PI controller in adapting to varying wind conditions. Furthermore, the analysis revealed the existence of chattering in the SMC used for MPPT.

The overall efficiency of the ANN controller-based MPPT surpasses that of the evaluated PI and SMC controllers. Its adaptability, absence of chattering phenomenon, faster response time and precise tracking collectively result in higher energy capture for wind energy generation. ANN controller-based MPPT exhibits remarkable speed in achieving a steady state operation, leading to faster recovery time when WS changes. Its ability to precisely track the MPP under varying wind conditions displays its accuracy and efficiency in power generation. The ANN-MPPT controller demonstrates itself as the most promising and effective solution for MPPT in WT systems. Its combination of speed, accuracy, robustness, and efficiency makes it a compelling

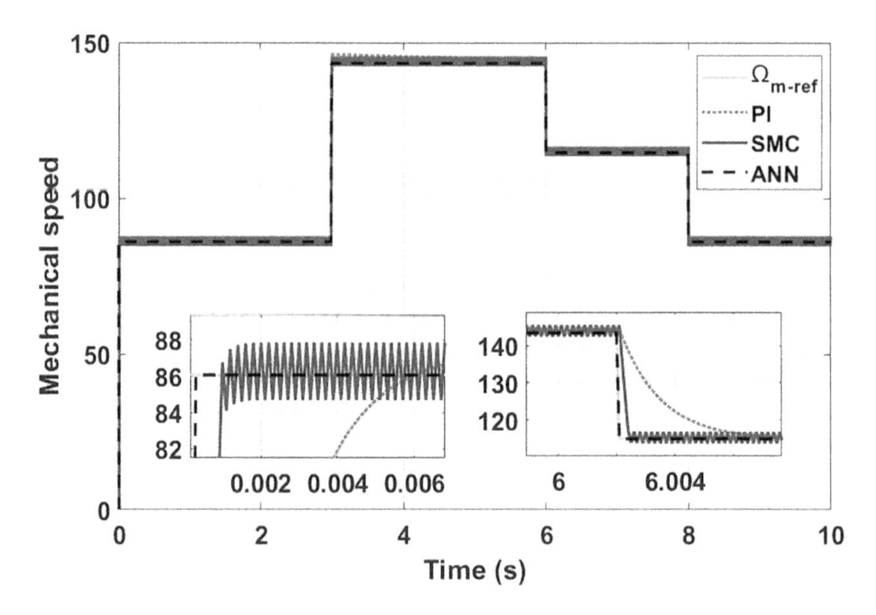

Figure 7.11 Mechanical speed Ωm for step change WS.

choice for optimizing power generation and maximizing the potential of wind energy resources.

7.4.2 Case 2: MPPT Performance under Random WS Variation

Figure 7.13 represents the random WS profile. To achieve the maximal extracted power from the supplied source, the speed ratio is set to λ_{opt} = 8.15, aligning with the maximum power coefficient C_{pmax} = 0.4794 (see Figure 7.14) for varying WS. Simulation results for MPPT using three suggested control algorithms (PI, SMC, and ANN), along with a mechanical speed control, as shown in Figure 7.15, indicate that all methods adeptly track the desired speed references for each WS value.

However, the SMC and PI controllers exhibit significant static errors, as shown in Figures 4.14 to 4.16. As depicted in Figure 4.14, the C_p response using the SMC methodology exhibits greater robustness compared to the PI controller. However, the SMC show a phenomenon of chattering, giving rise to various undesirable effects on both the quality of the generated electrical energy and the overall system performance.

Moreover, the ANN controller demonstrates the fastest response time and the most excellent tracking precision compared to the PI and SMC controllers. The higher error observed near the starting time (time zero) is due to the initially high damping at the turbine start-up.

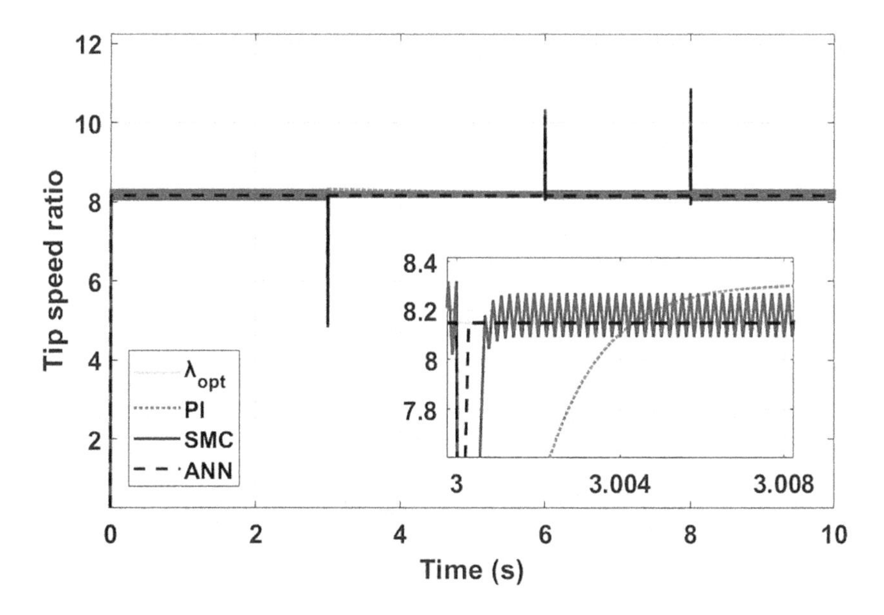

Figure 7.12 TSR λ for step WS.

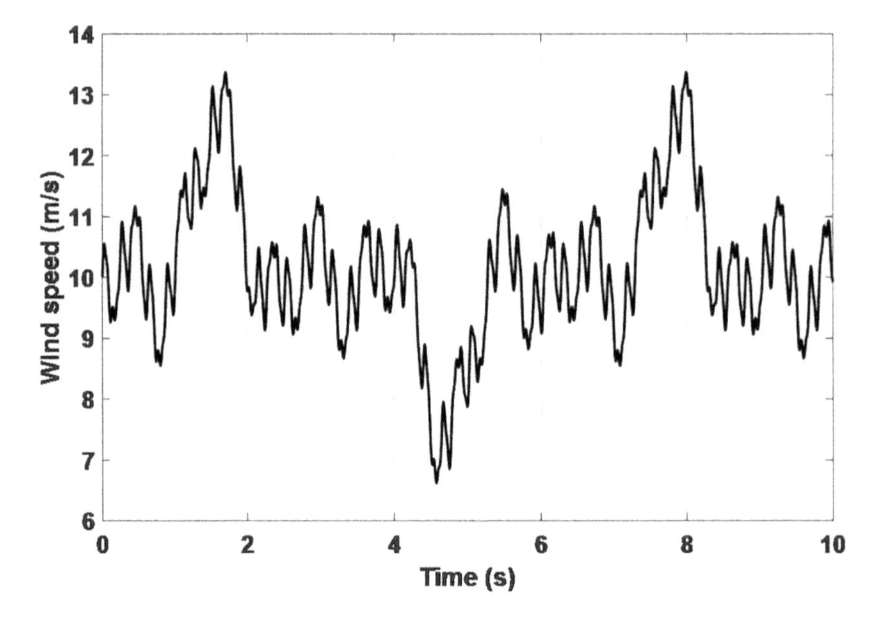

Figure 7.13 Random WS.

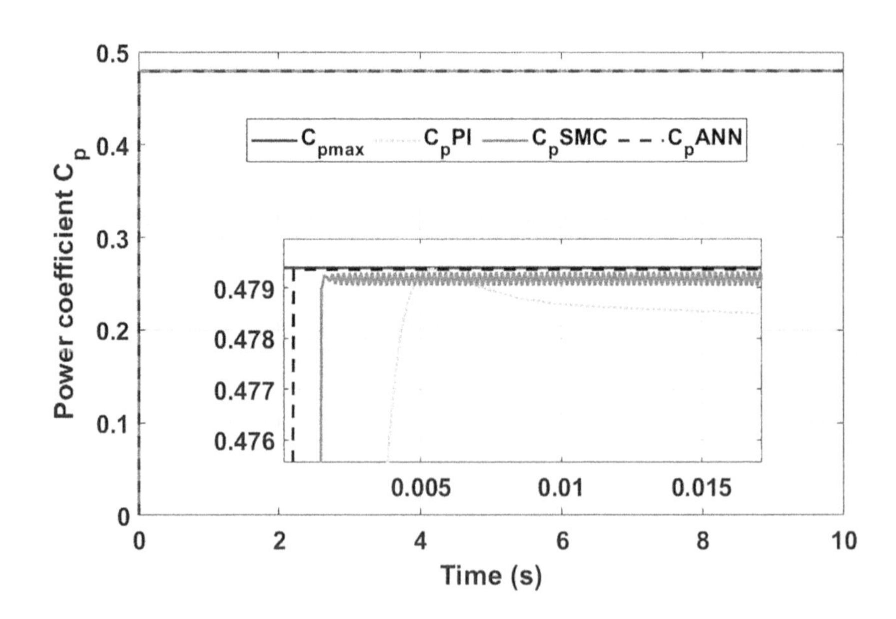

Figure 7.14 Power coefficient C_p for random WS.

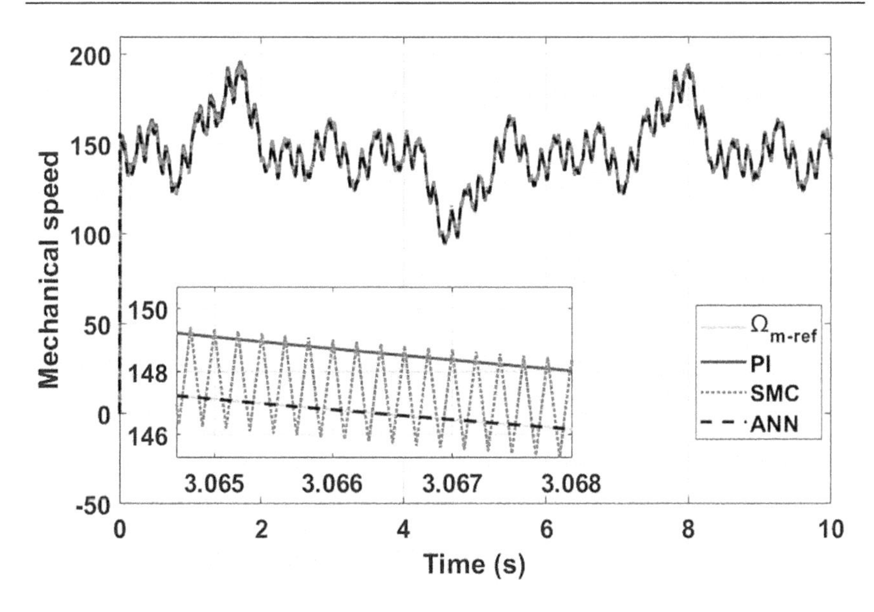

Figure 7.15 Mechanical speed Ωm for random WS.

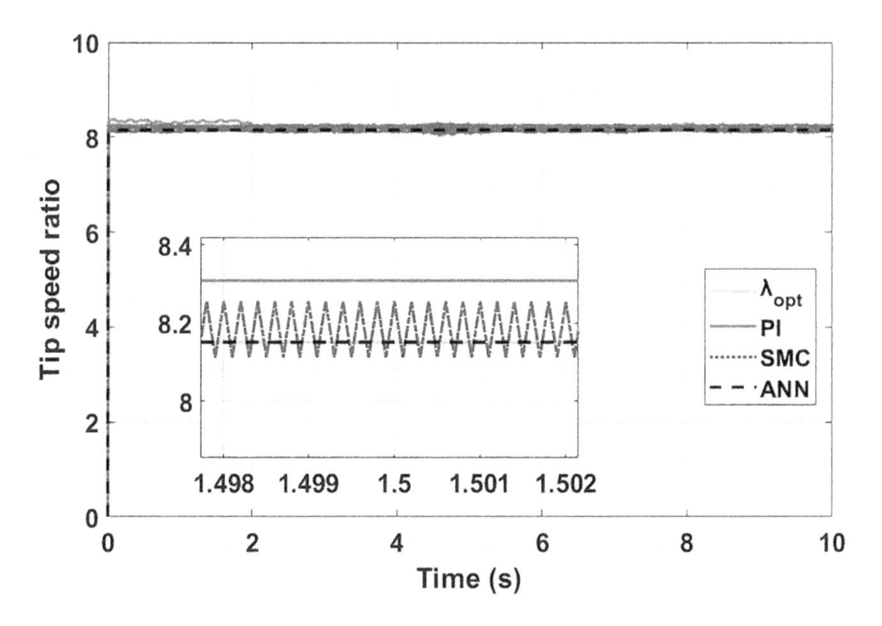

Figure 7.16 TSR λ for random WS.

Table 7.1 clearly shows that the ANN algorithm outperforms the PI and SMC controllers in MPPT for WTs. It achieves the lowest root mean square error (RMSE), mean squared error (MSE), and mean absolute error (MAE) values in both step and random WS scenarios, indicating its superior accuracy in tracking the MPP. With its accurate and efficient power point tracking capabilities, the ANN emerges as the most effective method among the compared algorithms. On the other hand, the PI algorithm shows relatively higher errors, especially in scenarios with dynamic WS changes.

Moreover, Table 7.2 presents a quantitative and qualitative synthesis comparing the three suggested controllers (PI, SMC, and ANN) in terms of precision, response time, and accuracy tracking. The table underscores substantial advancements achieved by the ANN controller, particularly in terms of response time optimization and precise tracking.

To evaluate the proposed ANN controller effectiveness in MPPT strategies, it is compared with other control methods from separate research papers. The comparison, shown in Table 7.3, considers various performance measurements. The results clearly indicate that the ANN controller achieves the fastest response time and maintains high precision and accuracy in tracking for the MPPT when compared to the other methods. These findings highlight the substantial benefits of leveraging advanced control techniques, such as the ANN, in improving the overall system efficiency and precision compared to conventional control methods. The superior results obtained with the ANN underscore the potential of neural network-based controllers for complex systems, where the precision and speed of response time are crucial factors for optimal system performance.

Table 7.1 RMSE, MSE and MAE Values by PI, SMC and ANN Algorithms for Two Scenarios of WS

	Case 1			*Case 2*		
MPPT method	RMSE	MSE	MAE	RMSE	MSE	MAE
PI	1.5715	2.4695	1.0406	2.6176	6.8518	1.7442
SMC	1.7378	3.0198	1.5900	2.1021	4.4189	1.5867
ANN	**0.7005**	**0.4906**	**0.0141**	**0.5271**	**0.2778**	**0.0420**

Table 7.2 Comparative Study between Suggested ANN-MPPT Algorithm and Other Work

Performance	*Response Time (ms)*	*Complexity*	*Precision*	*Set-Point Tracking*
PI	15	Very Low	Low	Low
SMC	1.9	Low	Chattering	Medium
ANN	**0.5**	**Medium**	**Excellent**	**Excellent**

Table 7.3 Comparative Study between Suggested ANN-MPPT Algorithm and Other Work

Reference Paper	*MPPT Method*	*Set-Point Tracking*	*Response Time (ms)*
[33]	PI	Low	200
	BSC	Good	5.2
	SMC	Good	3.5
	FLC	Very good	4
	ANNC	Excellent	2.4
[34]	Backstepping	Very good	5
[35]	OTC	Medium	24.88
[36]	ANNC	Excellent	4.5
[37]	ISMC	Good	280
Proposed method	**ANN**	**Excellent**	0.5

7.5 CONCLUSION

This chapter introduces ANN controller-based MPPT designed to optimize power extraction from VS WTs. Our objective is to achieve the maximum power output through peak power point tracking. The ANN training process involves input and target data, where the measured rotor speed and its reference constitute the inputs, and the torque reference serves as the output. The suggested method is implemented on a 3.5 kW VS WECS using MATLAB®/Simulink® and tested under various WS scenarios. Comparative analysis with PI and SMC controllers highlights the superiority of the ANN approach. It exhibits excellent power tracking and precision compared to PI and SMC controllers, even in the presence of significant WS fluctuations. This advantage extends across dynamic responses, reference tracking, precision, static errors, and robustness. Moreover, a comparative study with other methods of research work shows that the proposed ANN-based MPPT technique outperforms other methods in terms of response time; it achieves 0.5 ms response time while the other work ranging between 280 and 2.4 ms. The ANN controller holds the potential to enhance WT performance, effectively addressing challenges in current MPPT techniques. It excels in achieving remarkably low response time (as low as 0.0005s) and it consistently maintains the highest average C_p value, demonstrating its effectiveness in varying WS conditions.

REFERENCES

[1] Chainok, B., Tunyasrirut, S., Wangnipparnto, S., & Permpoonsinsup, W. (2017, March). Artificial neural network model for wind energy on urban building in Bangkok. In *2017 International Electrical Engineering Congress (iEECON)* (pp. 1–4). IEEE.

[2] Dolara, A., Gandelli, A., Grimaccia, F., Leva, S., & Mussetta, M. (2017, November). Weather-based machine learning technique for Day-Ahead wind power forecasting. In *2017 IEEE 6th International Conference on Renewable Energy Research and Applications (ICRERA)* (pp. 206–209). IEEE.
[3] Lin, W. M., & Hong, C. M. (2010). Intelligent approach to maximum power point tracking control strategy for variable-speed wind turbine generation system. *Energy*, 35(6), 2440–2447.
[4] Benamor, A., Benchouia, M. T., Srairi, K., & Benbouzid, M. E. H. (2019). A novel rooted tree optimization apply in the high order sliding mode control using super-twisting algorithm based on DTC scheme for DFIG. *International Journal of Electrical Power & Energy Systems*, 108, 293–302.
[5] Song, D., Tu, Y., Wang, L., et al. (2022). Coordinated optimization on energy capture and torque fluctuation of wind turbines via variable weight NMPC with fuzzy regulator. *Applied Energy*, 312, 312–118821.
[6] Tripathi, S. M., Tiwari, A. N., & Singh, D. (2015). Grid-integrated permanent magnet synchronous generator based wind energy conversion systems: A technology review. *Renewable and Sustainable Energy Reviews*, 51, 1288–1305. [CrossRef]
[7] Abdullah, M. A., Yatim, A. H. M., Tan, C. W., & Saidur, R. (2012). A review of maximum power point tracking algorithms for wind energy systems. *Renewable and Sustainable Energy Reviews*, 16, 3220–3227. [CrossRef] Energies 2021, 14, 6275 19 of 19.
[8] Kumar, D., & Chatterjee, K. (2016). A review of conventional and advanced MPPT algorithms for wind energy systems. *Renewable and Sustainable Energy Reviews*, 55, 957–970. [CrossRef]
[9] Apata, O., & Oyedokun, D. T. O. (2020). An overview of control techniques for wind turbine systems. *Scientific African*, 10, e00566. [CrossRef]
[10] Bendib, B., Belmili, H., & Krim, F. (2015). A survey of the most used MPPT methods: Conventional and advanced algorithms applied for photovoltaic systems. *Renewable and Sustainable Energy Reviews*, 45, 637–648. [CrossRef]
[11] Hosseini, S. H., Farakhor, A., & Haghighian, S. K. (2013, 28–30 November). Novel algorithm of maximum power point tracking (MPPT) for variable speed PMSG wind generation systems through model predictive control. In *Proceedings of the ELECO 2013–8th International Conference on Electrical and Electronics Engineering*, Bursa, Turkey, pp. 243–247. [CrossRef]
[12] Yu, K. N., & Liao, C. K. (2015). Applying novel fractional order incremental conductance algorithm to design and study the maximum power tracking of small wind power systems. *Journal of Applied Research and Technology*, 13, 238–244. [CrossRef]
[13] Hohm, D. P., & Ropp, M. E. (2003). Comparative study of maximum power point tracking algorithms. *Progress in Photovoltaics: Research and Applications*, 11, 47–62. [CrossRef]
[14] Cheng, M., & Zhu, Y. (2014). The state of the art of wind energy conversion systems and technologies: A review. *Energy Conversion and Management*, 88, 332–347. [CrossRef]
[15] Pagnini, L. C., Burlando, M., & Repetto, M. P. (2015). Experimental power curve of small-size wind turbines in turbulent urban environment. *Applied Energy*, 154, 112–121. [CrossRef]

[16] Lalouni, S., Rekioua, D., Idjdarene, K., & Tounzi, A. (2015). Maximum power point tracking based hybrid hill-climb search method applied to wind energy conversion system. *Electric Power Components and Systems*, 43(8–10), 1028–1038.
[17] Hannachi, M., Elbeji, O., & Benhamed, M., et al. (2021). Comparative study of four MPPT for a wind power system. *Wind Engineering*, 45, 1613–1622.
[18] Abdullah, M. A., Yatim, A. H. M., & Tan, C. W. (2014). An online optimum-relation-based maximum power point tracking algorithm for wind energy conversion system. In *Proceedings of the 2014 Australasian Universities Power Engineering Conference, AUPEC 2014 – Proceedings*, Perth, WA, 28 September–1 October.
[19] Tiwari, R., Krishnamurthy, K., Neelakandan, R. B., Padmanaban, S., & Wheeler, P. W. (2018). Neural network based maximum power point tracking control with quadratic boost converter for PMSG – wind energy conversion system. *Electronics*, 7(2), 20.
[20] Kumar, R., Agrawal, H. P., Shah, A., & Bansal, H. O. (2019). Maximum power point tracking in wind energy conversion system using radial basis function based neural network control strategy. *Sustainable Energy Technologies and Assessments*, 36, 100533.
[21] Assareh, E., & Biglari, M. (2015). A novel approach to capture the maximum power from variable speed wind turbines using PI controller, RBF neural network and GSA evolutionary algorithm. *Renewable and Sustainable Energy Reviews*, 51, 1023–1037.
[22] Ramesh Babu, N., & Arulmozhivarman, P. (2012). Forecasting of wind speed using artificial neural networks. *International Review on Modelling and Simulations*, 5(5), 2276–2280.
[23] Poultangari, I., Shahnazi, R., & Sheikhan, M. (2012). RBF neural network based PI pitch controller for a class of 5-MW wind turbines using particle swarm optimization algorithm. *ISA Transactions*, 51(5), 641–648.
[24] Thongam, J. S., Bouchard, P., Ezzaidi, H., & Ouhrouche, M. (2009, 8–10 July). Artificial neural network-based maximum power point tracking control for variable speed wind energy conversion systems. In *Proceedings of the IEEE International Conference on Control Applications*, St. Petersburg, Russia, pp. 1667–1671. [CrossRef]
[25] Ata, R. (2015). Artificial neural networks applications in wind energy systems: A review. *Renewable and Sustainable Energy Reviews*, 49, 534–562. [CrossRef]
[26] Hermassi, M., Krim, S., Kraiem, Y., Hajjaji, M. A., Alshammari, B. M., Alsaif, H., . . . & Guesmi, T. (2023). Design of vector control strategies based on fuzzy gain scheduling PID controllers for a grid-connected wind energy conversion system: Hardware FPGA-in-the-loop verification. *Electronics*, 12(6), 1419.
[27] Krim, Y., Abbes, D., Krim, S., & Mimouni, M. F. (2018). Classical vector, first-order sliding-mode and high-order sliding-mode control for a grid-connected variable-speed wind energy conversion system: A comparative study. *Wind Engineering*, 42(1), 16–37.
[28] Hermassi, M., Krim, S., Krim, Y., Hajjaji, M. A., Mtibaa, A., & Mimouni, M. F. (2022). Hardware FPGA implementation of an intelligent vector control technique of three-phase rectifier for wind turbine connected to the grid. In *2022 IEEE 9th International Conference on Sciences of Electronics, Technologies*

of Information and Telecommunications (SETIT), Hammamet, Tunisia, pp. 549–555, doi: 10.1109/SETIT54465.2022.9875830
[29] Hermassi, M., Krim, S., Krim, Y., Hajjaji, M. A., Mtibaa, A., & Mimouni, M. F. (2022). Xilinx-FPGA for real-time implementation of vector control strategies for a grid-connected variable-speed wind energy conversion system. In *2022 5th International Conference on Advanced Systems and Emergent Technologies (IC_ASET)*, Hammamet, Tunisia, pp. 49–54, doi: 10.1109/IC_ASET53395.2022.9765935
[30] Majout, B., Bossoufi, B., Bouderbala, M., Masud, M., Al-Amri, J. F., Taoussi, M., . . . & Karim, M. (2022). Improvement of PMSG-based wind energy conversion system using developed sliding mode control. *Energies*, 15(5), 1625.
[31] Li, G., & Shi, J. (2010). On comparing three artificial neural networks for wind speed forecasting. *Applied Energy*, 87(7), 2313–2320.
[32] Qiu, C., Yi, Y. K., Wang, M., & Yang, H. (2020). Coupling an artificial neuron network daylighting model and building energy simulation for vacuum photovoltaic glazing. *Applied Energy*, 263, 114624.
[33] Yessef, M., Bossoufi, B., Taoussi, M., Lagrioui, A., & Chojaa, H. (2022). Overview of control strategies for wind turbines: ANNC, FLC, SMC, BSC, and PI controllers. *Wind Engineering*, 46(6), 1820–1837.
[34] Nadour, M., Essadki, A., & Nasser, T. (2017). Comparative analysis between PI & backstepping control strategies of DFIG driven by wind turbine. *International Journal of Renewable Energy Research*, 7(3), 1307–1316.
[35] Nasiri, M., Milimonfared, J., & Fathi, S. H. (2014). Modeling, analysis and comparison of TSR and OTC methods for MPPT and power smoothing in permanent magnet synchronous generator-based wind turbines. *Energy Conversion and Management*, 86, 892–900.
[36] Boulkhrachef, O., Hadef, M., & Djerdir, A. (2021, January). Maximum power point tracking of a wind turbine based on artificial neural networks and fuzzy logic controllers. In *International Conference on Artificial Intelligence and its Applications* (pp. 100–111). Springer International Publishing.
[37]. Chojaa, H., Derouich, A., Chehaidia, S. E., Zamzoum, O., Taoussi, M., & Elouatouat, H. (2021). Integral sliding mode control for DFIG based WECS with MPPT based on artificial neural network under a real wind profile. *Energy Reports*, 7, 4809–4824.

Chapter 8

Enhancing Wind Energy Harnessing with an Intelligent MPPT Controller Using African Vulture Optimization Algorithm

K. Kathiravan and P. N. Rajnarayanan

8.1 INTRODUCTION

Progression in residential, commercial, and engineering sectors has been significantly influenced by energy. However, the necessity to investigate additional resources has arisen as a result of the rising need for energy [1]. Conventional energy sources have adverse effects on the atmosphere [2]. Renewable energy includes all power generated from the Sun, wind, water, and biological processes, whether directly or indirectly. These energies are infinite at a particular place and at a certain time [3]. Throughout the last decades, wind energy has proved to be the utmost dependable and proven energy source that is renewable. Due to its wide availability and non-depleting characteristics, wind power is a practical solution to the increasing need for clean and sustainable energy sources [4]. The kinetic energy generated by a wind turbine, that is connected to the prime mover through a gearbox, is transformed into electrical energy in a wind energy-producing system. The rotation of the generator's rotor begins right from the start, as it is directly linked to the wind turbine's rotor shaft. The wind turbine's mechanical power can be controlled within the range of wind speeds depending on cut-in and cut-out thresholds.

Permanent magnet synchronous generators (PMSG) or doubly-fed induction generators (DFIG) are the two main generators used by wind energy conversion systems (WECS). Various DFIG components, such as gearboxes, brushes, and slip rings, are not appropriate for certain applications. Due to the absence of these components, PMSG is very efficient, requires little upkeep, and is inexpensive. Additionally, it doesn't need any external excitation and can generate enormous torque at low speeds with minimal noise [5]. Recently, the WECS that relies on a PMSG has gained popularity, and it is used more often in big wind turbine generators for adjustable speed wind generation systems.

The wind turbines are categorized according to the wind speed, i.e. constant or changing. The economic viability of wind energy has increased significantly due to advancements in technology, notably since the advent of variable-speed WECS. Variable-speed wind turbines can adapt their

DOI: 10.1201/9781003462460-8

rotational speed to align with fluctuations in wind speed. Turbines with variable wind speeds are the only ones that can provide maximum power. To regulate the flow of power through this type of turbine, a power converter must be installed; this regulator is commonly referred to as a maximum power point tracker (MPPT). Despite being plentiful, wind energy fluctuates constantly due to fluctuations in wind speed. The MPPT tracks the highest power points, which determines how much power is obtained through WECS [6].

The aim is to create smart MPPT controllers tailored for wind turbine setups, enhancing the extraction of wind energy. The introduction of the AVOA is intended to elevate energy extraction efficiency.

The rest of the chapter is structured as follows: The conventional and intelligent MPPT controllers are described in Section 8.2. Section 8.3 focuses on the modeling and design of WECS components. The AVOA is detailed in Section 8.4. Section 8.5 presents the results and analysis. Conclusions are detailed in Section 8.6.

8.2 MPPT CONTROLLERS

8.2.1 Conventional Controllers

Among the explored algorithms for maximum power extraction, the three key control approaches studied so far include hill-climb search (HCS) control, power signal feedback (PSF), and tip-speed ratio (TSR) control [7]. To uphold the TSR at its ideal level, the TSR technique regulates the wind turbine's rate of rotation, which maximizes the amount of power extracted. To enable the system to derive the greatest power possible from the system, this approach necessitates the measurement or estimation of both wind and turbine speeds in addition to the knowledge of the optimal TSR of the turbine. Mastering the maximum power curve of a wind turbine and effectively employing its control methods are fundamental in PSF control. It is necessary to do simulations or offline tests on specific wind turbines to gain the maximum power curves. When employing the PSF approach, reference power is created either by applying the mechanical power equation for a wind turbine or using a maximum power curve that has already been recorded, with the input being either the wind speed or the rotor speed. The wind turbine's peak power is continuously sought after by the HCS control algorithm. Based on the system's operational point and the link between changes in power and speed, the tracking algorithm calculates the ideal signal to push the system to its maximum power [8].

In other words, direct power control (DPC) and indirect power control (IPC) are the two categories under which conventional procedures fall. The difference between these methods is that IPC methods track the mechanical output of the wind turbines whereas DPC methods track the maximum power of the electrical generator. Both TSR and PSF control algorithms are IPC techniques, and HCS is a DPC technique. DPC methods track the MPP by using a predetermined system curve. The most frequently employed HCS is regarded to be perturb and observe (P&O). Conversely, the tracking efficiency and speed of these conventional control algorithms are somewhat limited, and they are unable to extract the MPP, especially in circumstances where the wind speed varies rapidly [9].

Most of these simulations used conventional techniques such as P&O and modified P&O, whereas few researchers optimized either the proportional integral (PI) controller or the ANN-based controller using metaheuristic methodologies. There have been great efforts made in the simulation of the MPPT for wind energy generation systems. The metaheuristic optimization techniques are applied to overcome the problems associated with traditional methods. A flexible and reliable method for tackling challenging optimization issues is provided by metaheuristic optimization approaches. They require few presumptions, are relevant to a variety of scenarios, and can explore a variety of search spaces, adapt to different problem kinds, and handle real-world uncertainty. These methods excel at global search, strike a blend between exploration and exploitation, and are simple to parallelize for quick computing. Their flexibility, scalability, and success in practical applications highlight how important they are for taking on difficult optimization tasks [10]. A concise review of some of the metaheuristic optimization algorithms is presented here.

8.2.2 Intelligent Controllers

8.2.2.1 Artificial Neural Networks (ANN)

A computational model known as an artificial neural network (ANN) models biological neural networks found in the human brain after their structure and behavior [11]. It is a cornerstone of machine learning and the foundation for many deep learning techniques. An ANN is made up of layers of interconnected nodes, known as neurons, with three layers, such as input layer, an output layer, and one or more hidden layers. Every neuron layer uses weighted connections to process and send information, much like the synapses in the brain. Each neuron determines the weighted total of its inputs, processes the number through a function of activation, and sends the result to the following layer [12]. An ANN is trained by varying the link weights to minimize the discrepancy between the expected outputs and anticipated outputs of the system. Backpropagation is a widespread technique used in this process, which updates weights incrementally by propagating errors backward throughout the network using optimization methods such as gradient descent [13].

The WECS that is built using a DC/DC converter is suggested to use an ANN-based MPPT control technique. To maximize power extraction from the wind velocity, the suggested design employs a radial basis function network (RBFN). The study leads to the conclusion that in terms of maximum power extraction, an RBFN-based MPPT controller with a quadratic boost converter performs better than conventional techniques [14]. In comparison to the P&O technique, the implementation of the ANN and fuzzy logic controller (FLC) as a different MPPT algorithm led to better MPP tracking and quicker response times, and an increase in output power was achieved [15]. Through the training of varied weights for PMSG, sliding mode controller and ANN were used in several applications to augment the power output from the wind turbine [16].

Despite their effectiveness, ANNs have some drawbacks, such as the need for a lot of labelled data during training and their high computational requirements. However, their adaptability and capacity to extract complex features from data have produced groundbreaking advances in several sectors, propelling the expansion of AI applications across diverse fields.

8.2.2.2 Fuzzy Logic

Using fuzzy logic, which is a mathematical strategy, decision-making, and control systems may deal with uncertainty and imprecision. Fuzzy logic provides an alternative to traditional logic that can deal with circumstances when the lines separating true from false are unclear. It's especially helpful in systems that deal with ambiguous or complex input, where precise binary logic might not be appropriate, and allows machines to reason and choose in a manner like humans [17]. By permitting various levels of membership to a set, fuzzy logic expands on classical logic. Variables have several possible values outside just "true" or "false," ranging from 0 to 1, which indicate how much an element fits into a certain set. Key concepts include the degree of membership in a given set is indicated by the membership function, which transfers items of a discourse universe to a membership value between 0 and 1. To handle fuzzy sets and carry out reasoning, fuzzy logic makes use of set operations including union, intersection, and complement. Fuzzy rules have antecedents (conditions for the input) and consequents (activities for the outcome). From input values, they are utilized to deduce output values [18]. To enhance wind turbines' MPPT using PMSG, an artificially intelligent technique constructed on fuzzy logic controllers (FLC) is unveiled. The capability of FLC was compared with conventional TSR and PSF methods and FLC efficiency is higher than the other methods [19]. The accuracy of the MPPT approaches, however, could be impacted by uncontrollable factors like wind speed and air density, particularly during the wind speed minor oscillations. The Kalman filter employed to perform MPPT is used to calculate the uncertainty of the unexpected parameters. FLC is also used to regulate the generator's speed. FLC extracts maximum power even during minor oscillations [20].

Different interpretations may result from biased membership function implementation and sophisticated rule-based construction. Challenges include the complexity of the computation, the absence of formal procedures, and the difficulty of tweaking. Because of its intrinsic ambiguity and thin theoretical foundation, fuzzy logic can make analysis difficult.

8.2.2.3 Particle Swarm Optimization (PSO)

PSO, a widespread metaheuristic technique, is derived from the social conduct of birds [21]. PSO is built on the idea of modeling how several particles (agents) would behave as they moved across a multidimensional search space in pursuit of the best answer. Each particle serves as a potential answer to the optimization problem and modifies its location in the search space in response to its unique experience and the experiences of the particles around it. Every particle records its current velocity and location vectors. The velocity vector determines the particle's direction and speed within the search space, whereas the position vector represents a potential solution. Particles' motion is affected by both their current best-known position and that of their immediate surroundings [22].

To correct flaws present in the calculation of the generating circuit parameters and increase the efficiency of fixed-pitch DFIG wind turbines, an efficient PSO-based MPPT method was presented in [23]. A novel PSO-based MPPT is suggested for a free-standing self-excited induction generator (SEIG) functioning at changing wind speed and powering an induction motor linked to a centrifugal pump. The advantage of the suggested approach is that it does not necessitate an understanding of the parameters of the turbines of wind speed, air density, or both [24]. The duty cycle value is determined by the PSO technique based on the current and voltage measurements to control the boost converter. Simulation results showed that PSO, as opposed to conventional methods, guarantees consistent and reliable tracking of the highest power point, yielding more effective results [25].

The system's energy generation is enhanced in comparison to the outcomes of the traditional methods. A comparative study of the HCS hybrid solar/wind system and P&O approaches is carried out in [26]. Ant colony optimization (ACO) based MPPT controller is proposed in [27, 28]. The suggested approach combines proportional integral controllers and the ACO algorithm. It makes it possible to optimize the power captured and adjust the regulator's parameters according to the wind speed at any given moment. The grasshopper optimization algorithm (GOA) utilized in MPPT dynamically adjusts the duty cycle of the boost converter. This adaptation aims to optimize power extraction from a wind energy system operating under variable speed conditions. [29]. A technique called enhanced atom search optimization (EASO) that was created for WECS based on PMSG was used to achieve an ideal solution with high-quality and quick system response. The method offers a near-global optimal search capability to track the MPP. This method's potent global search capability, which facilitates quick response and high-quality optimal solutions, is one of its most promising advantages [30]. To replace the necessity for measuring instruments, get rid of system flaws, and reduce system size and expense, radial basis function-neural networks (RBF-NN) were recommended [31]. The RBF-NN technique performed better. The cuckoo search-based MPPT algorithm for tracking PV systems is discussed. Cuckoo search has several benefits, including a limited number of very efficient and quickly convergent parameter tuning steps [32]. To improve the PV system's convergence, speed, and precision, a genetic algorithm-based MPPT is presented [33]. The Archimedes optimization algorithm successfully addressed the shortcomings of the HCS approach regarding the effectiveness and speed of tracking the highest power point [34].

Despite the vast array of approaches utilized in modeling MPPT with wind energy systems, the actual use of metaheuristic approaches remains limited and warrants further consideration, despite their proven competence in achieving the highest power output point. To address the vacuum created by the use of the prior methodologies, a powerful metaheuristic technique, the African vulture optimization algorithm (AVOA) is advocated to construct MPPT equipped with a PMSG powered by a wind turbine. By allocating random and adaptive parameters, the convergence of AVOA can be managed. In addition, it accomplishes the phase blend between exploration and exploitation that allows the technique to reach the global optimum.

The AVOA is being proposed to build an MPPT controller integrated with a wind energy-producing system. The duty cycle of the boost converter is altered with the help of the suggested AVOA to boost the wind energy system's output power.

8.3 MODELING OF WECS

The wind turbine, PMSG, an uncontrolled rectifier, a DC-link capacitor, and a controller are among the parts of the small-scale WECS [35, 36]. Figure 8.1 depicts the design of the WECS taken into consideration in this work.

The AC output from the generator is changed to DC by a three-phase rectifier in the system, which also includes a wind turbine connected to a PMSG. The proposed AVOA-MPPT controller operates the boost converter after receiving the DC voltage and current from the three-phase rectifier. The proposed controller takes two inputs: electrical power and the mechanical speed of the wind turbine. The duty cycle is the output of the controller. The following is an explanation of the system model for each part of the built-in wind energy-producing system.

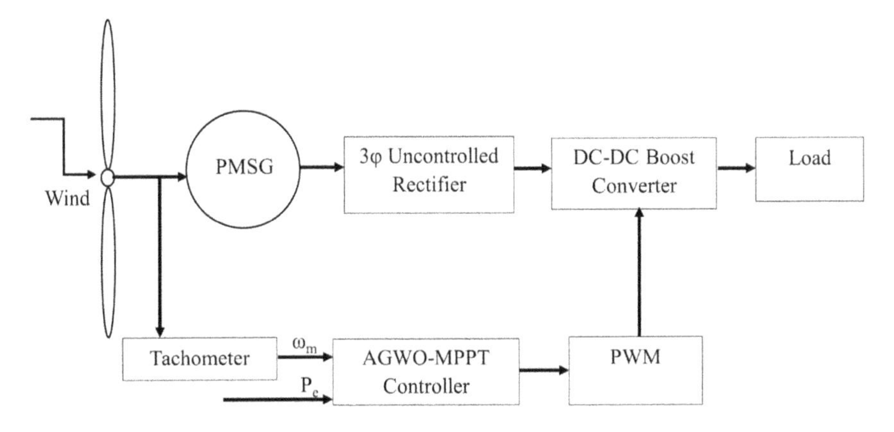

Figure 8.1 Design of WECS.

8.3.1 Wind Turbine

The functioning and effectiveness of a wind turbine are depicted through the modeling of its performance across diverse operational conditions. The description of the mechanical power yielded by the wind generators is as follows [37]:

$$P_m = \frac{1}{2}\rho A v^3 C_p(\lambda, \beta) \tag{8.1}$$

where,

- ρ air density (kg/m^3)
- A cross-section area of the wind turbine
- v speed of the wind (m/s)
- C_p power coefficient
- λ tip-speed ratio
- β pitch angle

The parameter C_p varies nonlinearly with the tip-speed ratio and pitch angle.

C_p is written as

$$C_p = C_p(\lambda,0) = C_p(\lambda) \tag{8.2}$$

In some theoretical analyses or basic models, the pitch angle might be assumed to be zero for simplification purposes. This assumption is often made because a pitch angle of zero means that the blades are not tilted or adjusted to capture more or less wind.

Equation 8.3 gives an expression of the rotational speed at the highest mechanical power.

$$\lambda^{opt} = \frac{R\omega_m^{opt}}{v} \tag{8.3}$$

where,

R radius of the blade

The maximum power is written as

$$P_m^{max} = \frac{1}{2} C_p^{max} \rho A v^3 \tag{8.4}$$

8.3.2 PMSG Model

PMSG is utilized in this work because it is effective and does not require a gearbox for operation [38]. PMSG does not require a gearbox in some applications because they have a high torque density and a wide speed range. This means that they can produce a lot of torque for their size and weight, and they can function across a broad spectrum of speeds without the need for a gearbox to change the speed ratio. The dynamic voltage equations along the axes of d and q are provided using,

$$u_d = -r_d i_d + \frac{d\psi_d}{dt} - \omega_e \psi_q \tag{8.5}$$

$$u_q = -r_q i_q + \frac{d\psi_q}{dt} - \omega_e \psi_d \tag{8.6}$$

where,

r_d & r_q stator resistance of axes of d and q

In the absence of rotor flux on the d and q axes, the currents along these axes are established by,

$$i_d = \frac{(\psi_{pm} - \psi_d)}{L_d} \tag{8.7}$$

$$i_q = -\frac{\psi_q}{L_q} \tag{8.8}$$

where,

L_d & L_q inductances of the axes

ψ_d & ψ_q flux linkages of the axes

The PMSG's electromagnetic power output is represented as

$$P_e = \frac{3}{2} \omega_e P \left[\psi_{pm} i_q - (L_d - L_q) i_d i_q \right] \tag{8.9}$$

The torques developed by PMSG is

$$T_e = \frac{P_e}{\omega_e} = \frac{3}{2}P\left[\psi_{pm}i_q - \left(L_d - L_q\right)i_d i_q\right] \tag{8.10}$$

where,

P pole pair number

8.3.3 DC-DC Boost Converter

The three-phase rectifier's output is supplied to the boost converter. Compared to other topologies, the boost converter offers a simpler and easier control because it just has one power-switching component. Next are the provided values for the converter's output current, output voltage, and power:

$$I_{out} = \frac{V_{in} * D}{R_{load}} \tag{8.11}$$

$$V_{out} = \frac{3V_{in}}{1-D} \tag{8.12}$$

$$P_{out} = V_{out} I_{out} \tag{8.13}$$

where,

V_{in} input voltage of the inverter

D duty cycle = $1 - \frac{V_{in}}{V_{out}}$

8.4 AFRICAN VULTURE OPTIMIZATION ALGORITHM

The African vulture optimization algorithm, which takes its cues from vultures' food-gathering habits, is the algorithm that this work proposes. AVOA takes its cues from the way African vultures hunt [39]. Predatory birds called vultures hunt sick or injured animals. The foraging tactics of vultures, which are renowned for their effective scavenging methods to locate carrion over wide lands, are imitated by this algorithm. Vultures aid in preventing the infection of the corpses [40]. The following criteria were used to model and simulate the dwelling arrangements and hunting habits of African vultures in order to create this algorithm [41]:

- Each of the N vultures in the African vulture population has a d-dimensional position space.

- In a natural setting, multiple vultures can segregate into two groups. The process initially evaluates the fitness of all potential options. The best vulture identified becomes the primary response, while the next best vulture is the secondary response. In each performance cycle, either the top two vultures are relocated or substituted by other vultures within the population. The groups can look for food in a variety of ways.
- The anti-hunger vultures are believed to be the most vulnerable and ravenous throughout the formulation stage. To find the best answers, others are edging away from them.

In AVOA, the best eagle in the group is chosen when four stages are completed: eagle starvation rate, exploration, and exploitation phases.

8.4.1 Phase 1: Choosing the Most Superior Vulture in a Group

The fitness of each member of the initial population is determined in this phase. The top vulture in the initial pack is termed the best, while the leading vulture in the second group is recognized as the finest. According to Equation 8.14, the remaining vultures adjust their positions towards the optimal solutions for both the primary and subsequent groups.

$$G(i) = \begin{Bmatrix} BestVulture_1 & if\ p_i = L_1 \\ BestVulture_2 & if\ p_i = L_2 \end{Bmatrix} \tag{8.14}$$

where,

Best_Vulture$_1$ & Best_Vulture$_2$ best vulture in first and second group
p_i probability
L_1 & L_2 random numbers in the rage [0 and 1]

The best vulture is chosen using a roulette wheel in the manner described here:

$$p_i = \frac{F_i}{\sum_{i=1}^{n} F_i} \tag{8.15}$$

where,

n total number of vultures
F_i fitness value

8.4.2 Phase 2: Rate of Vulture Starvation

Vultures have high amounts of energy after eating, allowing them to go further in pursuit of prey. Yet when they are in a state of hunger, vultures lack the strength for flying far and must instead go head-to-head for prey with more powerful vultures. Additionally, hungry vultures often get aggressive. This behaviour can be modelled as follows:

$$t = h \times \left(sin^{w}\left(\frac{\pi}{2} \times \frac{iter_i}{maxiter} \right) + cos\left(\frac{\pi}{2} \times \frac{iter_i}{maxiter} \right) - 1 \right) \tag{8.16}$$

$$F = (2r_1 + 1) \times z \times \left(1 - \frac{iter_i}{maxiter} \right) + t \tag{8.17}$$

where,

- h random value in [-2, 2]
- w fixed number, indicates exploration or exploitation phase
- F satiated vultures
- z random value in [-1, 1]
- r_1 random value in [0,1]

When the value of z falls below zero, the vulture goes hungry, but it is satisfied if the value of z rises.

Additionally, the AVOA has transitioned from the exploration stage to the exploitation stage by employing Equation 8.17. On the other hand, Equation 8.16 rises the probability of leaving from local optima, which improves optimizer performance while addressing complex problems.

As w rises, the likelihood of entering the exploratory stages in the final phase increases. Conversely, decreasing w diminishes the initiation of the exploratory phase. The vulture population is declining steadily, diminishing with each passing period. Once a value of |F| exceeds 1, vultures start to search for food in various locales, and the AVOA initiates the exploration phase. Vultures begin to forage around the solutions when AVOA triggers the exploitation phase, which occurs when the magnitude of |F| is less than 1.

8.4.3 Phase 3: Exploration

The ability to find weak, dying species in the wild and their good vision make vultures outstanding hunters. The task of finding food for vultures is really difficult. Vultures spend a considerable deal of time and effort carefully scrutinizing their living place and travel enormous distances in search of food. Vultures can arbitrarily scan a diversity of regions, and they can do this using one of two methods. Either strategy is chosen by a parameter named P_1. Which of the two approaches will be used depends on the value of this parameter, which needs to be decided prior to the search procedure and should vary from 0 to 1.

$$P_i(t+1) = \begin{cases} G(i) - |X - G(i) - P(i)| \times F P_1 \geq r_{P1} \\ G(i) - F + r_2 \times ((u_b - l_b) \times r_3 + l_b) P_1 < r_{P1} \end{cases} \tag{8.18}$$

Where,

G(i) one among the best vultures
X the distance, eagles fly to keep their prey safe from predators
r_2 & r_3 random value in [0,1]
l_b & u_b lower and upper bounds of search space

If r_3 value converges to unity, there will be a greater variety of methods and ability to search different regions of space.

8.4.4 Phase 4: Exploitation

The two techniques for the exploitation phase are chosen depending on the P_2 and P_3 parameter characteristics. The first parameter is used to select the strategy in the first stage, while the second parameter is used to assign the strategy in the second stage. When |F| ranges between 0.5 and 1, it signifies the initial phase of the exploitation stage, prompting the execution of two rotating flights and a siege-flight. The AVOA enters the competition for if $\lceil F \rceil$ is larger than or equal to 0.5. At this phase, the vultures possess ample energy and are typically content. When numerous vultures descend on a single food supply, there may be bloody conflicts for food. During these times, vultures with stronger bodies avoid feeding weaker ones. Contrarily, the lesser vultures gather around the stronger vultures and start fights in an attempt to drain them and capture their prey. The vultures then produce rotating flying to simulate spiral motion. The expression that follows can be used to describe the early stages of exploitation:

$$P_i(t+1) = \begin{Bmatrix} X - G(i) - P(i)(F + r_4) - (G(i) - P(i)) P_2 \geq r_{P2} \\ G(i) - G(i) \times \left(\frac{P(i)}{2\pi} \right) (r_5 \times \cos(P(i))) + r_6 \times sin(P(i)) P_2 < r_{P2} \end{Bmatrix} \quad (8.19)$$

where,

r_4, r_5 & r_6 random value in [0,1]

More vultures are drawn to the food source by the actions of two vultures, starting a violent siege war that is defined as the phase of exploitation's second stage.

$$P_i(t+1) = \begin{Bmatrix} 0.5 \left(Best_{Vulture1}(i) + Best_{Vulture2}(i) - \left(\frac{Best_{Vulture1}(i) \times P(i)}{Best_{Vulture1}(i) \times P(i)^2} + \frac{Best_{Vulture2}(i) \times P(i)}{2(i) \times P(i)^2} \right) \times F \right) P_3 \geq r_{P3} \\ G(i) - G(i) - P(i) \times F \times levy(G(i) - P(i)) P_3 < r_{P3} \end{Bmatrix} \quad (8.20)$$

where,

levy levy flight function

The techniques for the exploration and exploitation stages in AVOA are more transparent than those in other optimization systems. A balance between diversity and resonance is achieved in AVOA by using the two most best solutions as a representation of the two sets of vultures that are more powerful than the others, and it improves the performance of AVOA. Additionally, it executes the search space using a variety of diverse and varied techniques that were used in the early stages of the approach as well as the exploitation and exploration stages. The unique characteristic of AVOA, setting it apart from competitors and averting local optima, is its transition from exploration to exploitation. The strength of the AVOA lies in its low computing complexity and flexibility compared to other algorithms. Figure 8.2 shows the flow chart of AVOA.

8.5 RESULTS AND ANALYSIS

8.5.1 Initialization

This research work delves into two operational scenarios, considering both constant and fluctuating wind speeds. The fluctuation of C_p with λ for various wind turbine pitch angles is taken from [34]. λ_{opt} and C_p^{max} determine each curve's maximum power. Figure 8.3 displays the variance in wind turbine mechanical power and rotational speed. The suggested AVOA's proposed stages for designing the MPPT equipped with WECS are shown in Figure. 8.4.

A Simulink model for the proposed AVOA-based MPPT with WECS is developed. In Figure. 8.5, the Simulink model is displayed.

8.5.2 Parameters

The AVOA-MPPT controller receives both mechanical and electrical power, and its duty cycle is the output. The MOSFET duty cycle's lower and upper bands are designated as 0 and 1, respectively. The system specifications and parameters for AVOA are presented in Tables 8.1 and 8.2.

8.5.3 Analysis

8.5.3.1 Case 1: Constant Wind Speed

The initial setup examined in this research work is carried out with wind speed set at 12 m/s. The results are compared with the grey wolf optimization (GWO) algorithm. The results are given in Table 8.3; a variation of current, voltage, and maximum power is shown in Figures 8.6, 8.7, and 8.8 respectively.

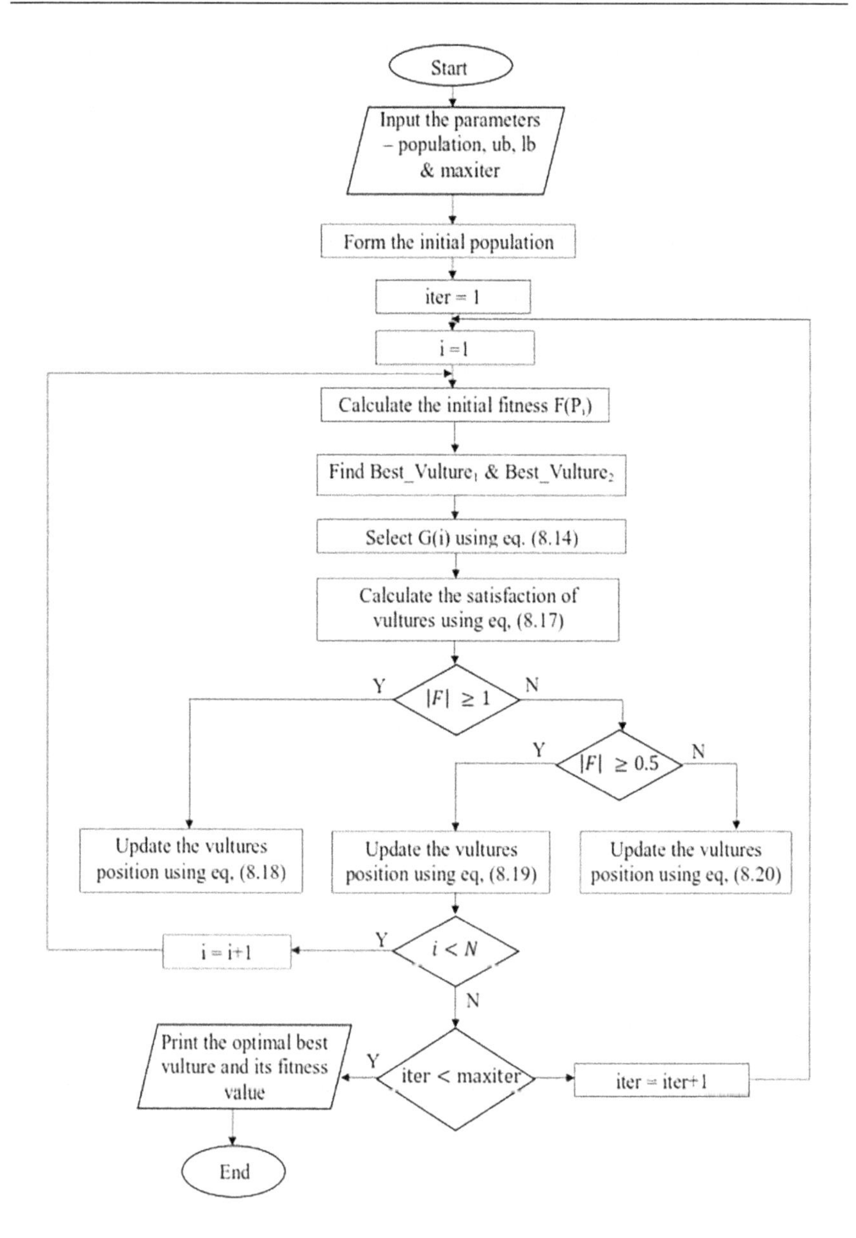

Figure 8.2 AVOA flowchart.

Table 8.3 demonstrates in clear terms that the suggested AVOA algorithm gives more voltage, current, and power than the GWO approach. The AVOA approach extracts 0.2% more power than GWO.

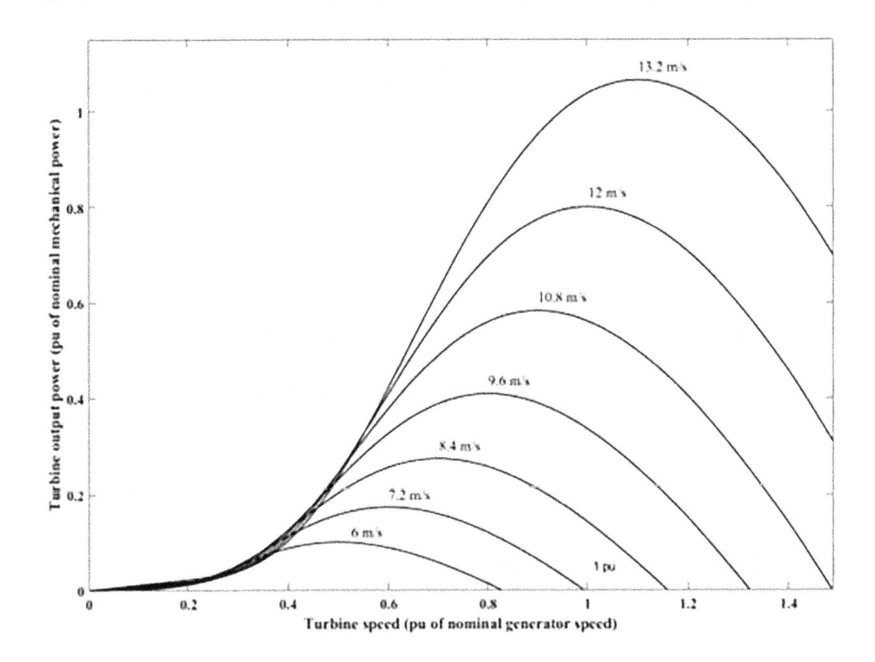

Figure 8.3 Mechanical Power vs wind speed.

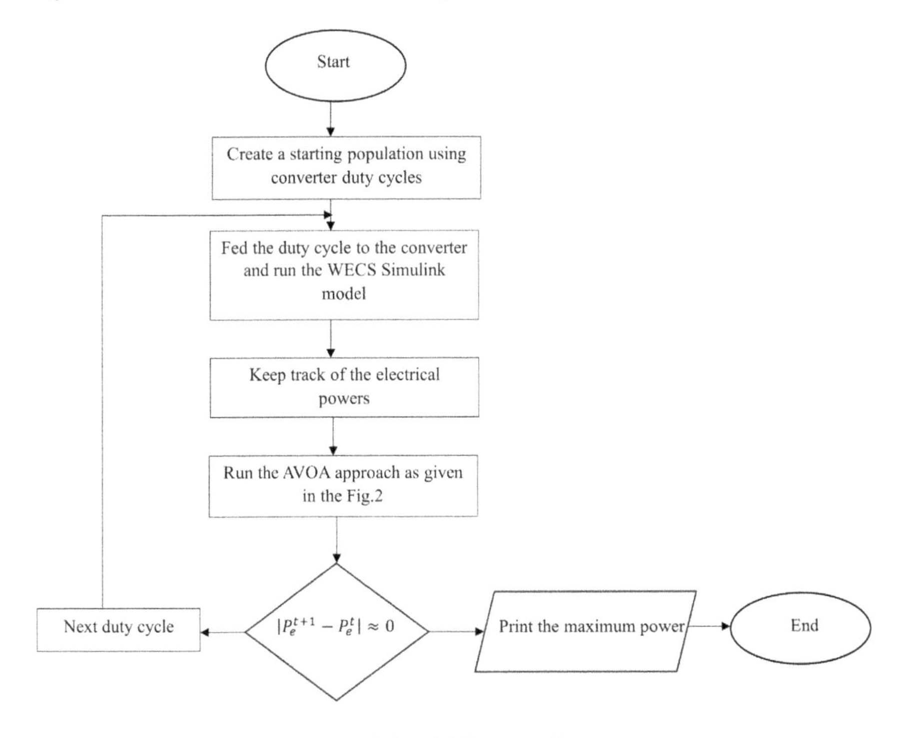

Figure 8.4 Steps for creating AVOA-MPPT controller.

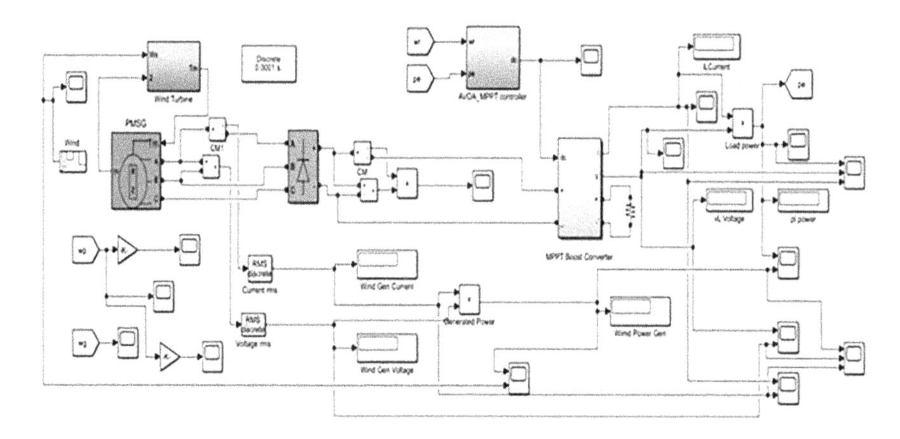

Figure 8.5 Proposed AVOA-MPPT controller's Simulink model.

8.5.3.2 Comparison of the Result

As a way to ensure the supremacy of the suggested AVOA algorithm, the results have been compared with other algorithms such as cuckoo search (CS), grasshopper optimization algorithm (GOA), electric charged particle optimization (ECPO), and Archimedes optimization algorithm (AOA) reported in [34]. Table 8.4 presents the obtained results.

It is clear from the justification given that the AVOA gives unquestionably superior performance compared to its algorithmic competitors, demonstrating an unmatched proficiency in harvesting maximum electrical power output. Significantly, the AVOA generated a peak power output of 104.9861W at a nominal duty cycle of 0.0893. Comparatively, the GWO algorithm, which was the next most effective method, reached its peak at 102.8751W while requiring a duty cycle of 0.0902.

The implications are clear: compared to its algorithmic competitors, AVOA successfully harnesses enhanced power from the converter, evidencing its exceptional capacity for crossing the global optimum. The resulting quantitative differences are expressed as follows:

- A remarkable 22.62% increase in power yield reflects AVOA's superiority against the CS paradigm.
- In a similar line, a 16.46% increase in power yield shows that AVOA outperforms ECPO.
- In terms of enhanced power, AVOA outperformed the GOA scheme by a margin of 0.43%.
- Furthermore, a 0.43% increase in power production demonstrates AVOA's superiority to the AOA.
- Not to mention, AVOA's slight but significant enhancement over GWO results in a 0.2% increase in power augmentation.

Table 8.1 Parameters of WECS

Parameter	*Value*	*Parameter*	*Value*
Wind Turbine		*PMSG*	
Nominal Power	250 w	Rated power	1 kW
Cut-in Speed	6 m/s	Rated Voltage	24 V
Cut-out speed	13.2 m/s	Stator resistance & inductance	0.05 Ω & 0.0085 H
β	0^0	Friction factor	0.011 kg.m^2
λ_{opt}	8.1	Damping factor	0.001889 N.m.s
Boost Converter		Pole Pairs	4
Capacitance (C_{in}, C_{out}, C_1 & C_2)	680 mF, 220 μF, 1.2 nF & 18 mF		
		Diode rectifier	
Load resistance	240 Ω	Forward voltage	0.8 V
n	10/4	Snubber resistance & capacitance	100 Ω & 0.1 μF
Inductance (L_m, L_{lk1} & L_{lk2})	19.14 mH, 0.011 mH & 0.031 mH	R_{on}	1 mΩ

Table 8.2 AVOA Parameters

Parameter	*Value*
Population	30
Maximum iterations	50
L_1 & L_2	0.8 & 0.2
P_1, P_2 & P_3	0.6, 0.4 & 0.6
w	2.5

Table 8.3 Comparison of GWO and AVOA

Parameter/Method	*GWO*	*Proposed AVOA*
P_m (w)	136.6310	137.1045
ω_m (rad/sec)	29.8721	30.3658
I_{MPP} (A)	0.6563	0.6613
V_{MPP} (V)	156.7654	158.7345
P_{max} (w)	102.8751	104.9861
Duty Cycle	0.0902	0.0893

Table 8.4 makes it obvious that the proposed AVOA approach extracts more power than the other reported approaches. The aforementioned comparisons create a convincing argument for the AVOA algorithm's superiority in this particular optimization environment. However, it is crucial to remember that the effectiveness of optimization algorithms depends on the

Table 8.4 Comparison of Results under Constant Wind Speed

Parameter/ Method	*CS*	*ECPO*	*GOA*	*AOA*	*GWO*	*Proposed AVOA*
P_m (w)	80.5449	95.9554	136.0344	136.0343	136.6310	137.1045
ω_m (rad/s)	17.5481	20.9039	29.3676	29.3676	29.8721	30.3658
I_{MPP} (A)	0.5145	0.57119	0.6493	0.6526	0.6563	0.6613
V_{MPP} (V)	123.5	137.0878	155.8435	156.6171	156.7654	158.7345
P_{max} (w)	63.5237	78.3044	101.1967	102.2039	102.8751	104.9861
Duty cycle	0.2000	0.1576	0.0975	0.0911	0.0902	0.0893

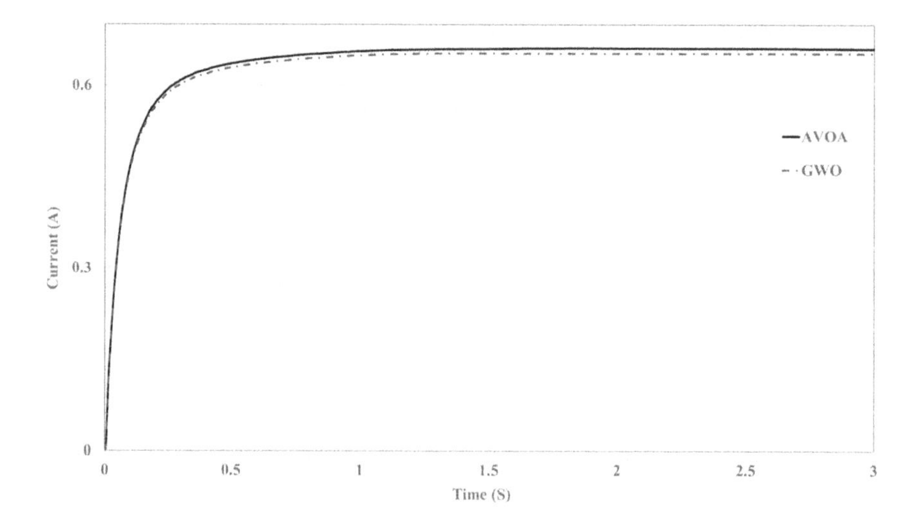

Figure 8.6 Variation of current.

particular issue domain, parameter settings, and a number of unrelated variables. The findings of this study augur well for the use of AVOA in orchestrating the converter MOSFET's duty cycle and maximizing electrical power output to its maximum.

8.5.3.3 Case 2: Variable Wind Speed

It is crucial to look at the designed MPPT in conjunction with the suggested AVOA when running the wind turbine in variable wind conditions. According to Figure. 8.9, there are five possible variations in wind speed.

Table 8.5 summarizes the results, which are compared with other reported algorithms [34].

Table 8.5 reveals that the suggested AVOA-MPPT algorithm notably exploited an excess of power from the wind producing apparatus

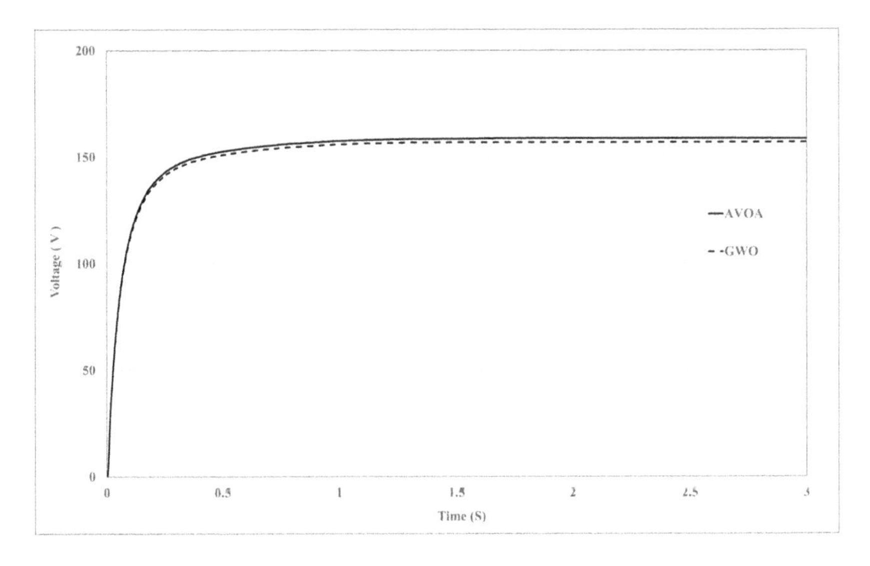

Figure 8.7 Variation of voltage.

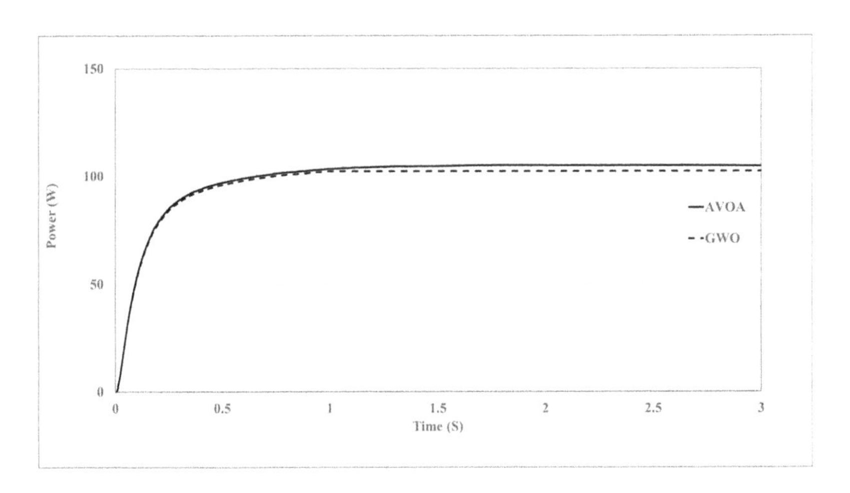

Figure 8.8 Variation of power.

during the second and fourth temporal intervals, quantified at magnitudes of 104.5330 and 104.6944 units, respectively. The subject algorithm clearly demonstrated a superior ability, surpassing its peers in the field of power optimization. In fact, even when dealing with the unpredictable speeds of the wind generation system, the AVOA algorithm obtains the best power outputs. This finding highlights the algorithm's remarkable effectiveness by demonstrating its capacity to take

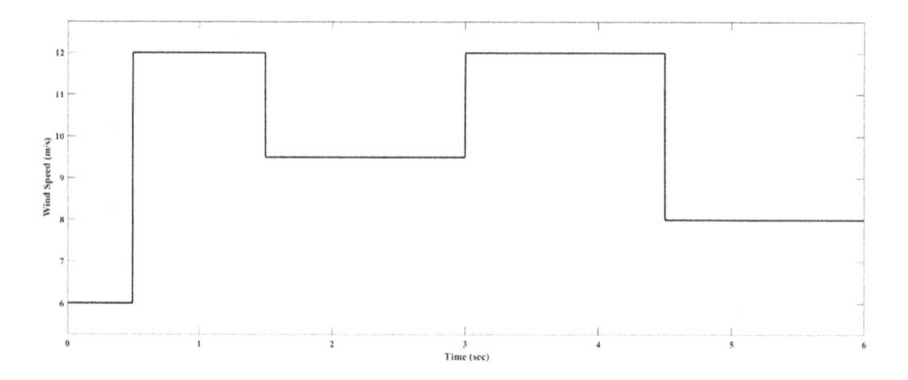

Figure 8.9 Time response of variable wind speed.

Table 8.5 Comparison of Results under Varying Wind Speeds

Parameter/Time Intervals		P_m *(w)*	ω_m *(rad/s)*	I_{MPP} *(A)*	V_{MPP} *(V)*	P_{max} *(w)*
0–0.5	GOA	8.603	7.747	0.159	37.68	6.217
	AOA	8.7363	7.6135	0.1631	39.136	6.3819
	GWO	8.9657	7.7218	0.1641	39.2678	6.4438
	AVOA	9.1198	7.8918	0.1650	39.7468	6.5599
0.5–1.5	GOA	135.9	29.3	0.651	156	102.13
	AOA	131.33	28.4347	0.6535	156.2965	102.136
	GWO	136.430	29.765	0.6533	156.3054	102.114
	AVOA	137.0189	30.1958	0.6608	158.2104	104.5430
1.5–3	GOA	55.83	18.65	0.414	99.44	41.2
	AOA	55.617	18.9225	0.4147	99.5323	41.02778
	GWO	56.7813	19.1025	0.4152	100.1923	41.5998
	AVOA	58.0586	19.6143	0.4196	101.084	42.4263
3–4.5	GOA	136.5	29.45	0.652	156.6	102.2
	AOA	135.6637	29.2949	0.6519	156.765	102.2028
	GWO	135.9854	29.7854	0.6528	156.4612	102.1378
	AVOA	137.1198	30.6834	0.6610	158.3879	104.6944
4.5–6	GOA	18.24	8.851	0.233	56.22	13.22
	AOA	28.2696	13.6855	0.2803	67.2717	18.8562
	GWO	28.7954	13.9547	0.2811	67.7543	19.0457
	AVOA	29.5106	14.1858	0.2836	68.321	19.38

advantage of transient changes in wind dynamics and reach the highest levels of power harvesting.

In conclusion, the experimental findings support the claim that the AVOA-MPPT algorithm holds a superior position over competing approaches. Its outstanding success in the duty of power extraction, particularly in the face of varying wind speeds, underlines its strength, and it is more suitable to optimize wind energy than other reported algorithms.

8.6 CONCLUSION AND FUTURE SCOPE

Finally, the findings reveal a fresh topic for investigation in the discipline of wind energy optimization through the deft use of the African vulture optimization algorithm (AVOA) within the context of MPPT management. Through the use of AVOA's inherent potential, this work intends to increase the effectiveness of wind energy generation systems. The subsequent AVOA approach, which is distinguished by its skillful modulation of the converter MOSFET's duty cycle, is a sign of increased power output in wind energy systems. The scope of the study includes both constant and fluctuating wind speeds in dynamic operational conditions. A thorough comparison is made between the outcomes of the AVOA and the well-known grey wolf optimization algorithm (GWO) as well as other algorithms reported in the literature. The suggested AVOA approach extracts more power than cuckoo search (CS), electric charged particle optimization (ECPO), grasshopper optimization algorithm (GOA), Archimedes optimization algorithms (AOA), and GWO approaches by 22.62%, 16.46%, 0.43%, 0.43%, and 0.2%. The results highlight the remarkable effectiveness of the AVOA-MPPT controller in improving wind energy systems, outperforming other optimizers. This method shows promise for solving optimization challenges in various renewable energy systems and the smart grid in the coming years.

Future advancements in intelligent MPPT controllers for wind turbine systems will focus on improving optimization algorithms, validating real-world implementations, hybrid control approaches, multi-objective optimization, grid integration, fault tolerance mechanisms, assessing environmental impact, coordinating control for wind farms, having adaptive learning capabilities, and utilizing hardware advancements for improved control precision.

REFERENCES

1. Umar DA, Alkawsi G, Jailani NLM, Alomari MA, Baashar Y, Alkahtani AA, Capretz LF, Tiong SK, "Evaluating the efficacy of intelligent methods for maximum power point tracking in wind energy harvesting systems", *Processes*, 2023, 11, 1420.
2. Nikolova S, Causevski A, Al-Salaymeh A, "Optimal operation of conventional power plants in power system with integrated renewable energy sources", *Energy Conversion and Management*, 2013, 65, 697–703.
3. Gaied H, Naoui M, Kraiem H, Goud BS, Flah A, Alghaythi ML, Kotb H, Ali SG, Aboras K, "Comparative analysis of MPPT techniques for enhancing a

wind energy conversion system", *Frontiers in Energy Research*, 2022, 10, 975134.
4. Roga S, Bardhan S, Kumar Y, Dubey SK, "Recent technology and challenges of wind energy generation: A review", *Sustainable Energy Technologies and Assessments*, 2022, 52, 102239.
5. Sonali R, Ramesh K, "Application of different MPPT algorithms for PMSG-based grid connected wind energy conversion system", *Engineering Research Express*, 2023, 5, 035021.
6. Abdullah MA, Yatim AHM, Tan CW, Saidur R, "A review of maximum power point tracking algorithms for wind energy systems", *Renewable and Sustainable Energy Reviews*, 2012, 16, 3220–3227.
7. Wang Q, Chang L, "An intelligent maximum power extraction algorithm for inverter-based variable speed wind turbine systems", *IEEE Transactions on Power Electronics*, 2004, 19(5), 1242–1249.
8. Thongam JS, Ouhrouche M, "MPPT control methods in wind energy conversion systems", *Fundamentals and Advanced Topics in Wind Power* (Ch. 15). InTech, 2011.
9. Ahmed J, Salam Z, "A modified P&O maximum power point tracking method with reduced steady-state oscillation and improved tracking efficiency", *IEEE Transactions on Sustainable Energy*, 2016, 7(4), 1506–1515.
10. Nassef AM, Abdelkaree MA, Maghrabi HM, Baroutaji A, "Review of metaheuristic optimization algorithms for power systems problems", *Sustainability*, 2023, 15, 9434.
11. McCulloch W, Pitts W, "A logical calculus of the ideas immanent in nervous activity" *Bulletin of Mathematical Biology*, 1943, 5, 115–133.
12. Zador A, "A critique of pure learning and what artificial neural networks can learn from animal brains", *Nature Communications*, 2019, 10, 3770.
13. Villegas-Mier CG, Rodriguez-Resendiz J, Álvarez-Alvarado JM, Rodriguez-Resendiz H, Herrera-Navarro AM, Rodríguez-Abreo O, "Artificial neural networks in MPPT algorithms for optimization of photovoltaic power systems: A review", *Micromachines*, 2021, 12, 1260.
14. Ramji T, Kumar K, Ramesh BN, Sanjeevikumar P, Patrick WW, "Neural network based maximum power point tracking control with quadratic boost converter for PMSG – Wind energy conversion system", *Electronics*, 2018, 7, 20.
15. Khan MJ, Kumar D, Narayan Y, Malik H, Márquez FPG, Muñoz CQG, "A novel artificial intelligence maximum power point tracking technique for integrated PV-WT-FC frameworks", *Energies*, 2022, 15, 3352.
16. Nirmal KA, Pradip KS, Suprava C, "MPPT based PMSG wind turbine system using Sliding Model Control (SMC) and Artificial Neural Network (ANN) based regression analysis", *IETE Journal of Research*, 2019, 68(3), 1652–1660. DOI: 10.1080/03772063.2019.1662336
17. Zadeh LA, "Fuzzy sets", *Information and Control*, 1965, 8(3), 338–353
18. Klir GJ, Yuan B, *Fuzzy Sets and Fuzzy Logic: Theory and Applications*, Prentice Hall, 1995.
19. Salem AA, Nour Aldin NA, Azmy AM, Abdellatif WSE, "A fuzzy logic-based MPPT technique for PMSG wind generation system", *International Journal of Renewable Energy Research*, 2019, 9, 4.

20. Honarbari A, Najafi-Shad S, Saffari Pour M, Ajarostaghi SSM, Hassannia A, "MPPT improvement for PMSG-based wind turbines using extended Kalman filter and fuzzy control system", *Energies*, 2021, 14, 7503.
21. Kennedy J, Eberhart R, "Particle swarm optimization", in *Proceedings of IEEE International Conference on Neural Networks*. IEEE, 1995, pp. 1942–1948.
22. Clerc M, Kennedy J, "The particle swarm – explosion, stability, and convergence in a multidimensional complex space", *IEEE Transactions on Evolutionary Computation*, 2002, 6(1), 58–73.
23. Sompracha C, Jayaweera D, Tricoli P, "Particle swarm optimization technique to improve energy efficiency of doubly-fed induction generators for wind turbines", *Journal of Engineering*, 2019, 4890–4895.
24. Mohamed Ali Zeddini, Remus Pusca, Anis Sakly, Faouzi Mimouni, "PSO-based MPPT control of wind-driven Self-Excited Induction Generator for pumping system", *Renewable Energy*, 2016, 95.
25. El Asiiaout H, El Ougli A, Tidhafi B, "MPPT using PSO technique comparing to fuzzy logic and P&O algorithms for wind energy conversion system", *WSEAS Transactions on System and Control*, 2022, 17, 305–313.
26. Kumar AV Pavan, Parimi Alivelu M, Uma Rao K, "Implementation of MPPT control using fuzzy logic in solar-wind hybrid power system", *IEEE International Conference on Signal Processing, Informatics, Communication and Energy Systems (SPICES)*, 2015, 1–5.
27. Sabrina T, Cherif L, Kamal YT, Karima B, "A new MPPT controller based on the Ant colony optimization algorithm for Photovoltaic systems under partial shading conditions", *Applied Soft Computing*, 2017, 58, 465–479.
28. Mokhtari Y, Rekioua D, "High performance of maximum power point tracking using ant colony algorithm in wind turbine", *Renewable Energy*, 2018, 126, 1055–1063.
29. Fathy A, El-baksawi O, "Grasshopper optimization algorithm for extracting maximum power from wind turbine installed in Al-Jouf region", *Journal of Renewable and Sustainable Energy*, 2019, 11(3), 033303.
30. He X, We P, Gong X, Meng X, Shan D, Zhu J, "Enhanced atom search optimization based optimal control parameter tunning of PMSG for MPPT", *Energy Engineering: Journal of the Association of Energy Engineers*, 2022, 119, 145–161.
31. Kumar R, Agrawal HP, Shah A, Bansal HO, "Maximum power point tracking in wind energy conversion system using radial basis function based neural network control strategy", *Sustainable Energy Technologies and Assessments*, 2019, 36, 100533.
32. Mossad MI, Osama Abed al-Raouf M, Al-Ahmar MA, Banakher FA, "Maximum power point tracking of PV system based cuckoo search algorithm: Review and comparison", *Special Issue on Emerging and Renewable Energy: Generation and Automation, Energy Procedia*, 2019, 162, 117–126.
33. Slimane H, Jean-Paul G, Fateh K, "Real-time genetic algorithms-based MPPT: Study and comparison (theoretical an experimental) with conventional methods", *Energies*, 2018, 11, 459.
34. Fathy A, Alharbi AG, Alshammari S, Hasanien HM, "Archimedes optimization algorithm based maximum power point tracker for wind energy generation system", *Ain Shams Engineering Journal*, 2022, 13, 101548.

35. Priyadarshi N, Ramachandaramurthy VK, Padmanaban S, "An ant colony optimized MPPT for standalone hybrid PV-wind power system with single Cuk converter", *Energies*, 2019, 12(1), 167.
36. Oussama M, Abdelghani C, Lakhdar C, "Efficiency and robustness of type-2 fractional fuzzy PID design using salps swarm algorithm for a wind turbine control under uncertainty", *ISA Transactions*, 2022, 125, 72–84.
37. Kazmi SMR, Goto H, Guo HJ, Ichinokura O, "Review and critical analysis of the research papers published till date on maximum power point tacking in wind energy conversion system", *IEEE Energy Conversion Congress and Exposition (ECCE)*, 2010, 4075–4082.
38. Tripathi SM, Tiwari AN, Singh D, "Grid-integrated permanent magnet synchronous generator-based wind energy conversion systems: A technology review", *Renewable and Sustainable Energy Reviews*, 2015, 51, 1288–1305.
39. Salah B, Hasanien HM, Ghali FMA, Alsayed YM, Abdel Aleem SHE, El-Shahat A, "African vulture optimization-based optimal control strategy for voltage control of islanded DC microgrids", *Sustainability*, 2022, 14, 11800.
40. Abdollahzadeh B, Gharehchopogh FS, Mirjalili S, "African vultures optimization algorithm: A new nature-inspired metaheuristic algorithm for global optimization problems", *Computers and Electrical Engineering*, 2021, 158, 107408.
41. Mohana A, Ahmed F, Dalia Y, Hegazy R, "Optimal reconfiguration of shaded PV based system using African vultures optimization approach", *Alexandria Engineering Journal*, 2022, 61, 12159–12185.

Chapter 9

Renewable Energy Powered Switched Reluctance Motor for Marine Propulsion System Using Soft Computing Techniques

G. Jegadeeswari, D. Lakshmi, and B. Kirubadurai

9.1 INTRODUCTION

In an effort to create "green ships," research institutes in a number of maritime nations are looking at fuel cells, wind, solar, nuclear, and biomass energy. It is frequently regarded as a kind of environmentally friendly ship design with practically infinite possibilities for advancement. Solar ships harness the energy of the sun via the use of solar panels and storage batteries to generate electricity for propulsion.

The maritime sector is essential to international trade and provides island people with a vital lifeline. Research on the construction of all-electric ships (AES) has gained traction due to growing concerns about environmental protection and energy utilization. Electric motor drives, which comprise the majority of AES technologies, must provide high power density, wide speed operating range, high controllability, high efficiency, and maintenance-free operation. The majority of these objectives can be met by the switching reluctance (SR) motor drive.

The SRM is seen as a potential solution for maritime applications because of advantages such as high-speed operation and inherent fault tolerance. In this study, SRM is powered by PV to power the ship's propeller. In order to work correctly, the SRM requires a steady and continuous supply of electricity. Many converters with efficient controllers are available in the literature for this purpose. In this study, a SEPIC converter is used to increase the DC voltage from the PV system. ANFIS-based MPPT algorithm is described to deliver the input signal to the PWM generator attached to the converter. The procedure for choosing a particular controller is covered in-depth in this chapter. Furthermore, an analysis was conducted on the efficiency of the PV system, SEPIC converter, and energy storage system that powered the SR motor.

In the proposed method, a mechanically coupled propeller is driven by a PV-powered SR motor. An SR motor is used to counteract the abrupt surge in speed and torque experienced by the electric propulsion system based on induction motors. The photovoltaic (PV) system needs a sufficient maximum power point tracking (MPPT) technique in order to enhance the PV output

DOI: 10.1201/9781003462460-9

power. Due to the unpredictable nature of temperature and solar radiation, traditional MPPT systems have low efficiency and severe oscillations at maximum power point. The suggested adaptive (ANFIS)-based MPPT can handle these problems. This study examined an adaptive (ANFIS)-based maximum power point tracking of a photovoltaic system constructed with a single ended primary inductance converter (SEPIC). For MPPT, choosing the right converter is essential. In terms of efficiency, SEPIC converters perform better than other converters.

A SEPIC converter is used to enhance the low-level, fluctuating DC voltage generated by the PV system. The ANFIS-MPPT receives the PV cell's real voltage V_PV and current I_PV as input and generates a control signal via effective error elimination. When this control signal is sent into the PWM generator, it creates PWM pulses that control the SEPIC converter's switching operations, resulting in a regulated DC output voltage. This output voltage is delivered to the switched reluctance motor via the bridge resonance converter topology, which energizes the phase windings of the SRM, in order to regulate the torque and speed of the 8/6 SRM.

Because of its many advantages, including low cost, high durability, high fault tolerance capabilities, and high efficiency, switched resistor motors have recently gained popularity in changeable speed applications. For this reason, SRM is regarded as an effective motor in maritime propulsion systems driven by renewable energy sources. However, the large torque ripple of a switching reluctance motor presents a significant obstacle in its development. The torque ripple causes vibration and noise, which lowers its performance. One of the main areas of research for switching reluctance motors (SRM) is torque-ripple reduction.

Soft computing techniques can provide approximations for difficult or unsolvable complex problems. Soft computing techniques provide an efficient way to deal with nonlinear problems. They provide affordable solutions to challenging real-world problems for which there isn't a solid computer solution. The key concern in an appropriate control approach becomes the minimization of the torque ripple when smooth control is needed at low speeds. In this study, torque ripples in switching reluctance motors are reduced by using soft computing techniques like fuzzy logic controllers (FLC). The performance of a traditional proportional integral (PI) controller is also taken into consideration for comparison's sake. To power the ship propeller in this project, a four phase, eight-switched reluctance motor is employed.

9.2 LITERATURE REVIEW

The rising prominence of fuel leakage based ecological disasters in addition to the increased intensity of carbon emissions from fossil fuel powered ships have contributed to contamination of the marine environment.

Thereby, with the aim of curbing the yearly emission of greenhouse gases and lowering the carbon intensity, the Marine Environmental Protection Committee (MEPC) has advocated at its 72nd conference to increase the energy efficiency design index criteria for ships. Moreover, the International Convention for the Prevention of Pollution has put forward several stringent rules for curtailing marine pollution. The Paris Agreement in addition to the Kyoto Protocol further emphasizes the need to cut down carbon emission in order to meet the goal of keeping the global warming temperature below 2°C. Consequently, it is crucial to facilitate the development of green ships and intensify research on deployment of renewable energy technology on marine applications as reported by Abdel-Rahim & Wang [1], Abid Ali et al. [2], Al Nabulsi & Dhaouadi [3], and Ahmad & Narayanan [4]. As demonstrated by Alassi & Massoud [5], the development of semiconductor switching devices for use in high power drives and the modern application of electric propulsion started in the 1980s. These developments allowed for complete control over the speed of the thrusters and propellers, which in turn allowed for the simplification of the mechanical structure. But the main argument in favour of using electric propulsion in commercial ship applications is the possibility for fuel savings when compared to comparable mechanical choices. The authors claim that since converter technology developed so swiftly, AC drives eventually replaced DC drives. Following AC drives, pod propulsion was introduced, in which an electric motor was placed directly on the propeller shaft within a submerged, 360-degree steerable pod. It is shown that using electric propulsion can result in lower installation costs and space requirements, better manoeuvrability, and enhanced efficiency.

Owing to the disadvantages of traditional power generation, including cost, pollution to the environment, and fuel scarcity, research on the efficient use of non-conventional energy sources has been concentrated in recent decades. Solar energy is a significant renewable energy source among the numerous non-conventional sources. The ocean and land surfaces of the globe absorb solar energy, which is then put to various uses. Longer term advantages are associated with sustainable solar energy systems. Photovoltaic, concentrated solar power, water treatment, and latent heat storage are all used in active solar approaches. As suggested by Albert Alexander et al. [6], the decreasing cost of PV modules, free energy source, low maintenance, and noiseless operation make solar PV systems essential for meeting energy generating needs.

The stand-alone single-stage solar photovoltaic (SPV) three-phase sensorless induction motor drive for a water pumping application was covered by Singh et al. [7] with the aid of the proposed four-switch inverter. At different degrees of sun irradiation, it was both simulated and experimentally executed. The adaptive incremental conductance technique is used to calculate the SPV reference DC-link voltage, while direct vector control is utilized to manage the induction motor drive's speed. Furthermore, the

DC-link voltage is maintained constant by the use of a reference voltage control, even in the presence of fluctuating amounts of radiation. Directly from the MATLAB® environment, the four-switch inverter's control signal is generated using DSPACE DS1104. The overall performance of an induction motor drive has been shown to be well-suited for both starting and steady state operations, despite fluctuations in solar insolation levels.

Narendra et al. has developed and studied a novel design for a solar-powered water pumping system that uses the grid to help offer an uninterrupted water supply in 2022 [8]. It has been demonstrated that the developed grid-side VSC control functions well in every operating mode. When solar power is not enough to pump the water at its full capacity, the grid provides the energy required to run the system at installed capacity. For the bulk of the system's operational life, the SRM drive has been seen to operate at a fundamental frequency.

Sharma et al. [9] have provided comprehensive classifications and a brief description of different PV-fed drives based on applications, types of conversion stages, different converters, and control methods in order to achieve the required output for the efficient operation of electric drives. Power conditioning equipment and DC motors powered directly by photovoltaic cells are used. The study also discusses other types of DC-DC power converters, such as buck, boost, buck-boost, cuk, Sepic, and Luo converters, for different purposes.

To overcome the constraints and provide higher stability margins, attenuation at the resonance frequency, and increased resilience, the authors proposed employing a second-order controller rather than a PI controller. A full analysis of the resulting closed-loop stability was reported by G. Jegadeeswari et al. in 2020 [10], and the experimental findings are presented to support the controller design.

In 2022 [11], A. Ibrahim et al. discussed the static synchronous compensator (STATCOM) for sustaining continuous operation of renewable energy systems (RESs) during three-phase faults occurring at the point of common coupling (PCC) between the RESs and the grid, as well as improving the performance of a hybrid energy system containing SRG-based WECS and photovoltaics (PVs) against wind gusts. A hybrid model combining STATCOM and grid integration is simulated using MATLAB®/Simulink®. The whale optimization algorithm (WOA) and particle swarm optimization (PSO) are utilized to optimize the scheduling of the two PI controller settings for regulating the STATCOM. The voltage at the point of common coupling (PCC) is adjusted by adjusting the reactive power flow between the STATCOM and the hybrid system. The system's reactive power can be enabled during wind gusts and fault events by using the WOA to control the STATCOM rather than the PSO. In a fair comparison of the quantitative and qualitative performance of the STATCOM-PI controllers optimized by WOA and PSO, WOA outperforms PSO in terms of dynamic performance.

9.3 PROPOSED METHODOLOGY

The suggested study effort describes a PV-fed 4 SRM that is utilized to power the ship's propeller, which is physically attached to the motor. Intelligent controllers and various metaheuristic optimization techniques are used in SRM to reduce torque ripples and ensure smooth operation. The abrupt increase in speed and torque of traditional induction motors is optimally corrected by substituting an induction motor with an SR motor for marine propulsion systems. Due to the use of a hysteresis current control technique, the speed of the SRM does not drop even when a load is applied.

PV produces a low-level DC voltage with fluctuations, which is improved by using a high gain SEPIC converter. The output of a solar system has been managed and maximized by the use of high gain SEPIC. Furthermore, the DC-DC converter is regulated using an ANFIS-based MPPT algorithm. The ANFIS-MPPT accepts the true voltage VPV and current IPV of the PV cell as input and provides a control signal through effective error elimination. When this control signal is fed into the PWM generator, it generates PWM pulses that govern the switching processes of the SEPIC converter, resulting in a regulated DC output voltage. Furthermore, a battery system with a bidirectional DC-DC converter is used to store extra energy from PV.

The SEPIC converter output drives the phase windings of the switched reluctance motor via the (n+1) semiconductor and (n+1) diode converter. Furthermore, a PI controller, intelligent controllers, and an optimization technique for adjusting PI are used to control the speed of SRM. The controller gets the actual speed of the motor N_act as well as the reference speed N_ref, resulting in an error signal. This error is transmitted to the controllers, which generate a control signal by employing effective error elimination. When this control signal is fed into the PWM generator, it generates PWM pulses that control the switching operations of the (n+1) semiconductor and (n+1) diode converter, also known as bridge resonance converter, thereby energizing the phase windings of the SRM. Figure 9.1 depicts the overall block diagram of the suggested systems.

9.4 RESULTS AND DISCUSSION

This section examines the speed control of an 8/6 SR motor powered by a PV-fed SEPIC converter. The converter is controlled using an ANFIS-based MPPT algorithm, and the resulting performance is examined. For PV systems, several factors like as temperature, radiation, voltage, and current are examined, and the corresponding waveforms are obtained. Table 9.1 lists the specifications of the solar panels and SEPIC converter, followed by the outputs simulated using MATLAB® platforms.

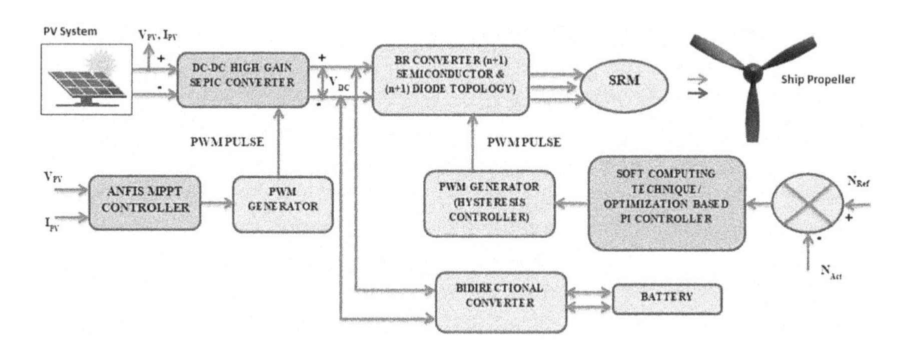

Figure 9.1 Overall block diagram of proposed system.

Table 9.1 Specifications of SEPIC Converter

Solar Panel's Parameters	
Parameters	*Ratings*
Operating Voltage	16.8 *V*
No. of series cells	36
Maximum Voltage	1000 *V*
No. of panels	10
Temperature Range	−40 *to* +85°C
Operating Current	5.8 *A*
SEPIC Converter Ratings	
Diodes	*MCD95*
L_1, L_2	0.576 *mH*
C_1, C_2	22μF
Switching Frequency	10 KHz

Figure 9.2(a) shows that the solar panel temperature waveform is progressively maintained constant at 250° Celsius and then suddenly rises to 350° Celsius after 0.2 seconds, which is then maintained constant.

Figure 9.2(b) shows the solar panel irradiation waveform. The irradiation progressively kept constant at 800 w/sq.m and a rapid spike to 1000 w/sq.m at 0.2 s, which is further maintained constant.

The voltage progressively remains constant at 125 V and suddenly rises to 149 V after 0.2 s; following which it remains constant at 149 V. Figure 9.2(c) depicts the voltage received from the solar panel.

It is observed that there is a significant rise in the current at the beginning. However, the current progressively declines and settles down at 60 A and again increases with a spike in current of 70 A at 0.2 s and remains constant after 0.2 s. Figure 9.2(d) depicts the current obtained from the solar panel.

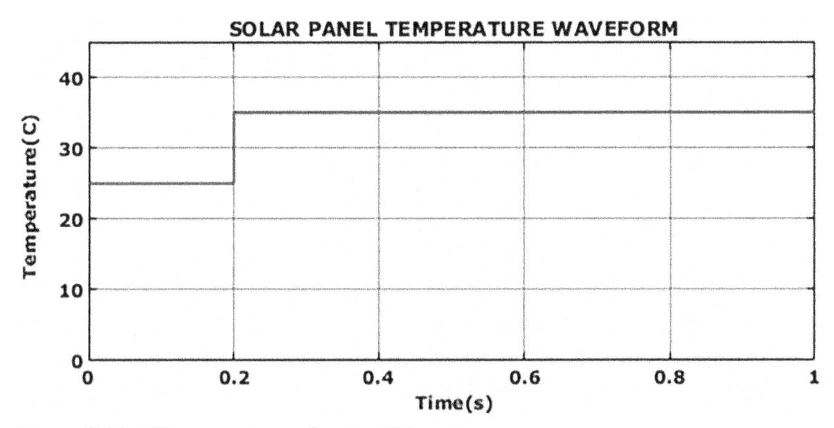

Figure 9.2(a) Temperature of solar PV system.

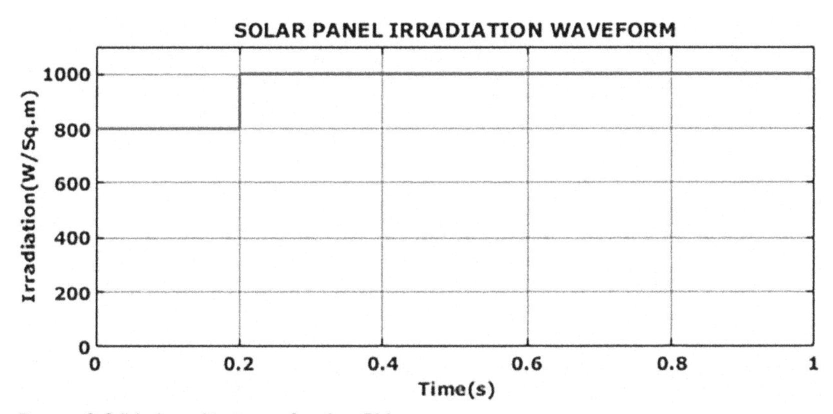

Figure 9.2(b) Irradiation of solar PV system.

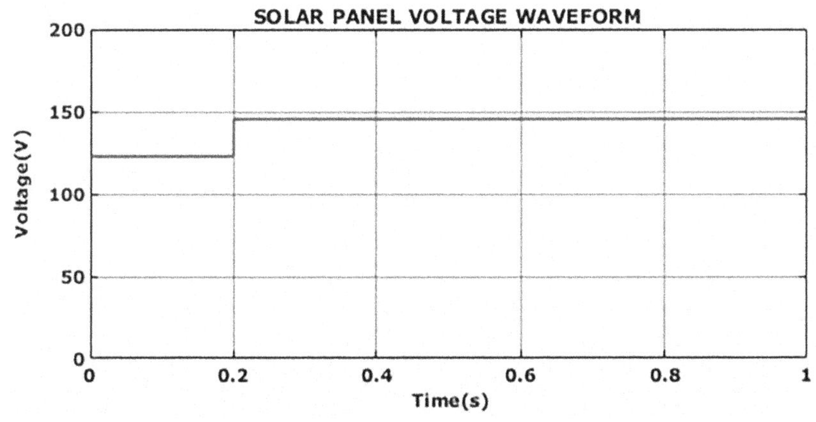

Figure 9.2(c) Voltage waveform of solar PV system.

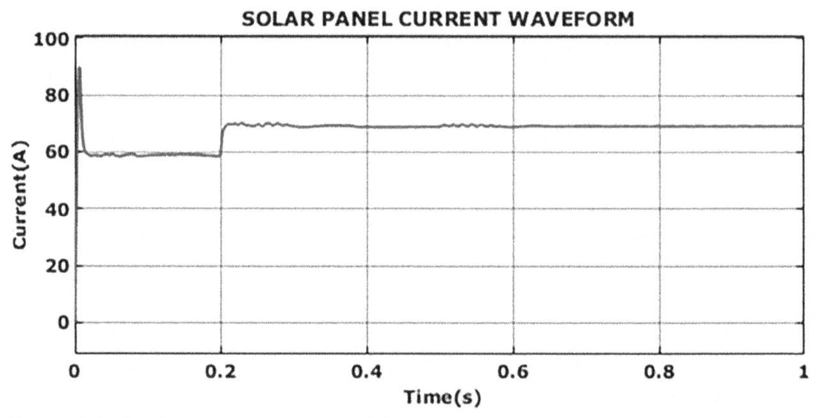

Figure 9.2(d) Current Waveform of Solar PV system.

The solar side converter output voltage and current waveforms for the SEPIC converter fed ANFIS-based MPPT algorithm are shown in Figures 9.3(a) and (b).

The output voltage of the SEPIC converter is around 450 V with a solar panel input of 140 V. The usage of a high gain SEPIC DC-DC converter boosts the output of the PV panel. The conversion efficiency is approximately 97%. The SEPIC converter output drives the phase windings of the switched reluctance motor via the (n+1) semiconductor and (n+1) diode converter. The voltage received from the converter is shown in Figure 9.3(a).

The output current achieved by the SEPIC converter is around 22 A. Figure 9.3(b) shows the current received from the converter.

Figures 9.4(a), (b), and (c) depict the output voltage and current waveforms, as well as the steady of charge (SOC) waveforms, for a battery-powered energy storage system that employs a bidirectional converter.

Figure 9.4(a) indicates that the output voltage from the battery is approximately 24 V. Figure 9.4(b) illustrates the output current from the battery, which is around 1.2 A.

Figure 9.4(c) shows that the battery's SOC is approximately 80%.

The speed control of four phase 8/6 SRM fed renewable energy based marine propulsion system are analyzed. The speed control of the motor is carried out by traditional and soft computing techniques and the corresponding performances are analyzed. Various time domain specifications like rise, peak and settling time, maximum overshoot, steady state error, and torque ripple are analyzed, and the related flux, current, and torque waveforms are attained. The specification of SRM is listed in Table 9.2 following which the outputs simulated using MATLAB platforms are provided.

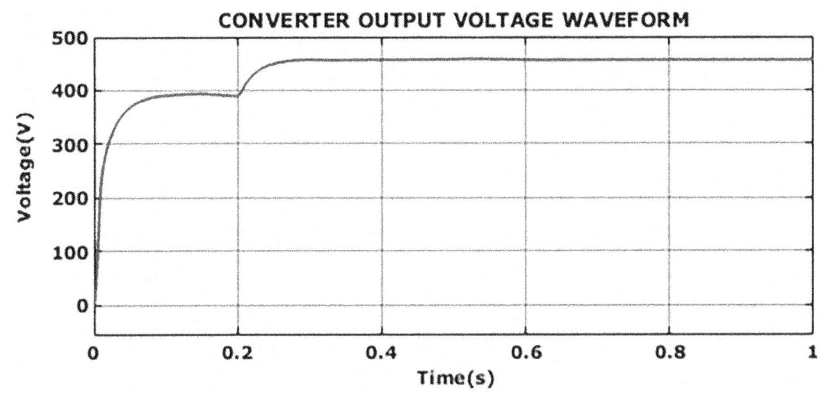

Figure 9.3(a) Voltage waveform of SEPIC converter.

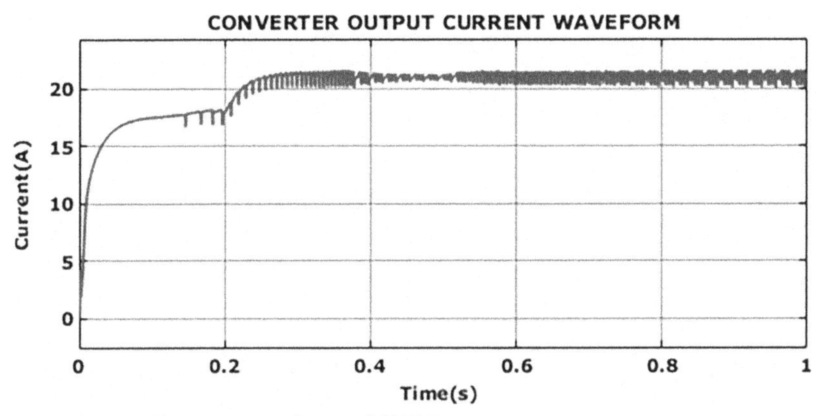

Figure 9.3(b) Current waveform of SEPIC converter.

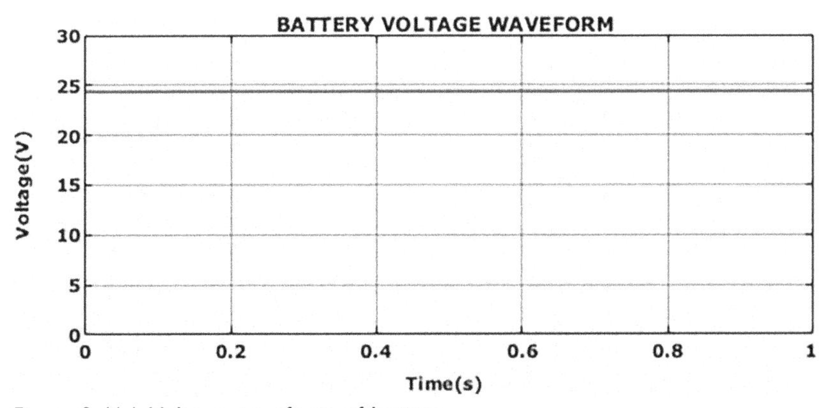

Figure 9.4(a) Voltage waveform of battery.

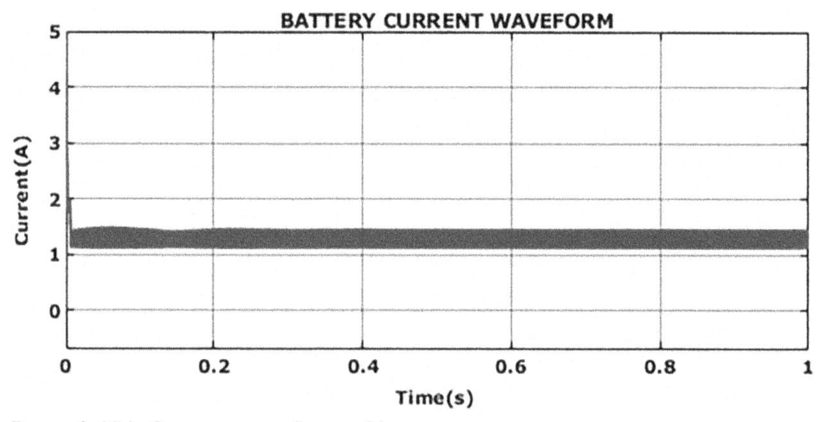

Figure 9.4(b) Current waveform of battery.

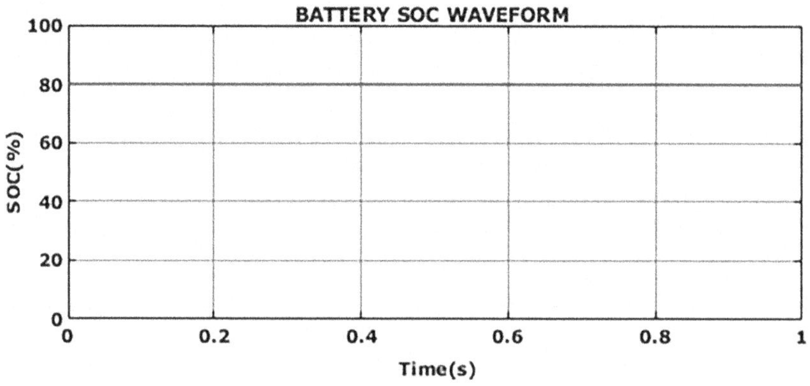

Figure 9.4(c) SOC waveform of battery.

9.4.1 Speed Regulation of SR Motor Using Proportional Integral Controller (PI)

With a view to study the effectiveness of speed control of SRM using PI controller, simulation studies are performed. The corresponding flux, current and torque waveforms for time duration of 0 *to* 1 *s* are presented in Figures 9.5(a)–(d) respectively.

From the flux waveforms, it can be observed that the flux increases and gradually decrease at 0.5 s and continues to maintain a constant value of 0.082 *Wb*, at 1 *NM*. Figure 9.5(a) shows the flux behaviour of SRM using PI.

From the following waveforms it is observed that initially the current rises rapidly and drops down. After mild fluctuations, the current is settled to a constant value of 10 *A*, at 1 *NM*. Figure 9.5(b) shows the current behaviour of SRM using PI.

Table 9.2 SRM Specification

Number of Phase & stator/rotor poles	4 ϕ & 8/6
Rated power	10 *KW*
Rated voltage	600 *V*
Speed	6000 *rpm*
Saturated aligned inductance	$0.15e^{-3}H$
Unaligned inductance	$0.67e^{-3}H$
Aligned inductance	$23.6e^{-3}H$
PWM rectifier switch	IP module (1200 *V*, 25 *A*)
PWM generator	FPGA Spartan 6E controller
Switch	IGBT
Driver Circuit	TL250

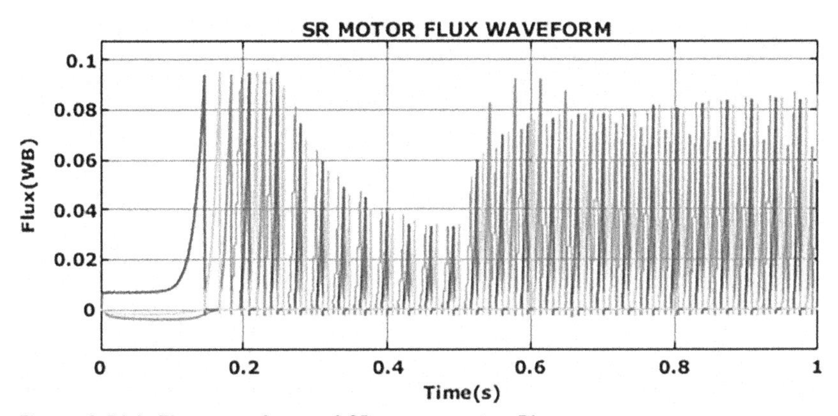

Figure 9.5(a) Flux waveform of SR motor using PI.

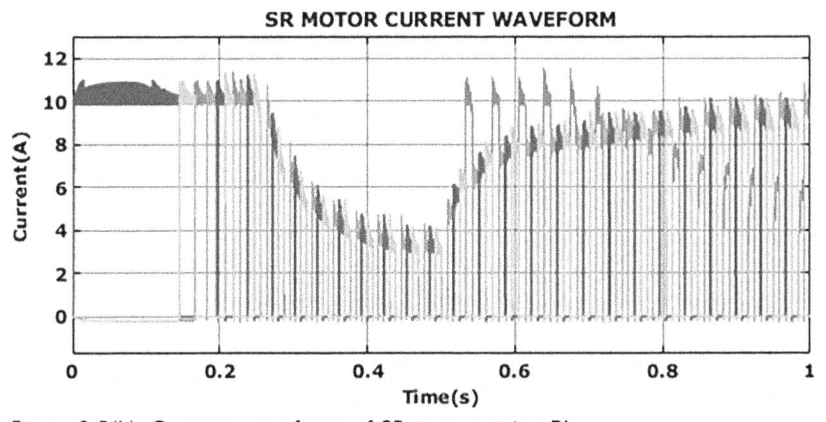

Figure 9.5(b) Current waveform of SR motor using PI.

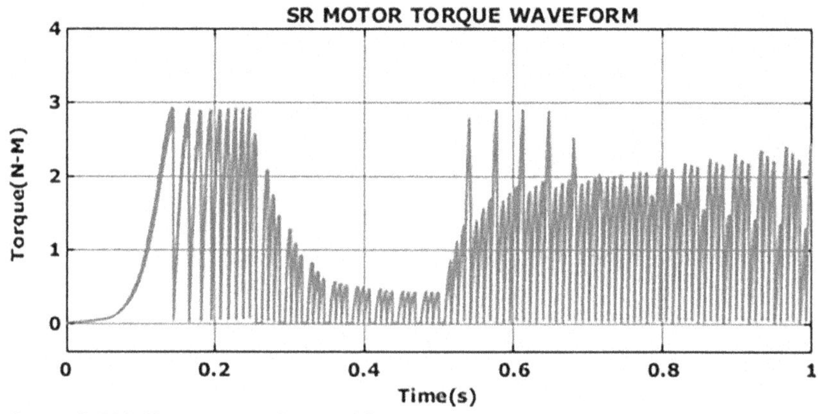

Figure 9.5(c) Torque waveform of SR motor using PI.

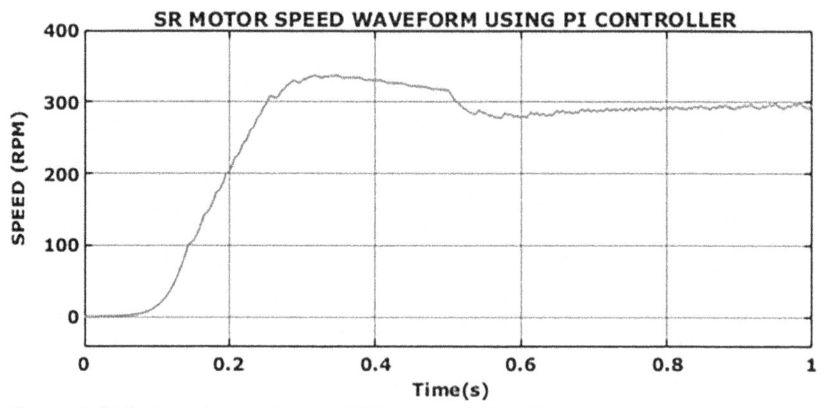

Figure 9.5(d) Speed waveform of SR motor using PI.

From the following torque waveforms, it can be observed that the torque undergoes a spike and drops down. After mild fluctuations it settles to a constant value of 2.5 *NM*. Figure 9.5(c) shows the torque behaviour of SRM using PI.

Simulation studies are expanded to study the speed behaviour of SRM, and it can be seen that the value of speed attempted at 1 NM is 300 rpm. The response clearly demonstrates that larger maximum deviation, longer response time, and longer duration of oscillation.

The simulation was run numerous times to fine-tune the PI controller to present the optimal parameter values. Table 9.3 shows the tuned parameters coefficient of the PI controller that resulted in the speed response graph shown in Figure 9.5(d).

The performance and robustness of PI controller in terms of time domain parameters is presented in Table 9.4.

Table 9.3 PI Controller Parameters

Controller Parameters	*Parameters Values*
P	0. 0035358
I	0. 0064514

Table 9.4 Performance and Robustness of PI Controller

Parameter	*Tuned*
Rise time (t_{rl}) in Sec	0.32
Peak Time (tp) in Sec	0.34
Overshoot in %	0.5
Settling Time (t_s) in Sec	0.9
Maximum Peak Overshoot	0.56
Steady State error	0.82
Integral Absolute error	16.81
Torque ripple in %	89.90

The torque ripple in the PI controller is found to be quite significant, at 89.90%. To some extent, several soft computing techniques are used for SRM speed control, as shown in the following results.

9.4.2 Speed Regulation of SR Motor Using Fuzzy Logic Controller (FLC)

With a view to study the effectiveness of speed control of SRM using FLC controller, simulation studies are performed. The corresponding flux, current and torque waveforms for time duration of 0 *to* 1 *s* are presented in Figures 9.6(a)–(d) respectively.

From the flux waveforms, it is noted that the value of flux is 0.082 *Wb*, at 1 *NM*. It can be observed that the control approach adapted is capable of maintaining the flux constant at the attained values. Figure 9.6(a) shows the flux behaviour of SRM using FLC.

From the following waveforms it is noted that initially the current rises rapidly and drops down. After mild fluctuations, the current is settled to a constant value of 11 *A*, at 1 *NM*. It can be observed that the control approach adapted is capable of maintaining the current constant at the attained values. Figure 9.6 (b) shows the current behaviour of SRM using FLC.

From the following torque waveforms it can be noted that the torque undergoes a spike and drops down. After mild fluctuations it settles to a constant value of 2.8 *NM*. It can be observed that the control approach adapted is capable of maintaining the current constant at the attained values. Figure 9.6(c) shows the torque behaviour of SRM using FLC.

Simulation studies are expanded to study the speed behaviour of SRM, and it can be seen that the speed obtained is 300 rpm at 1 NM. The response clearly demonstrates that larger maximum deviation, longer response time,

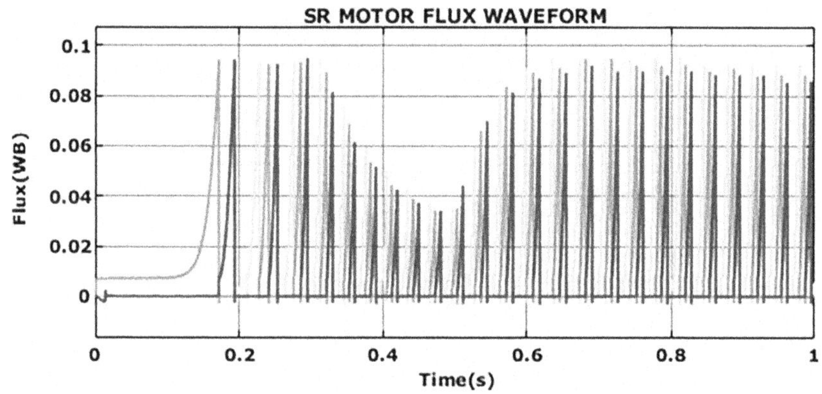

Figure 9.6(a) Flux waveform of SR motor using FLC.

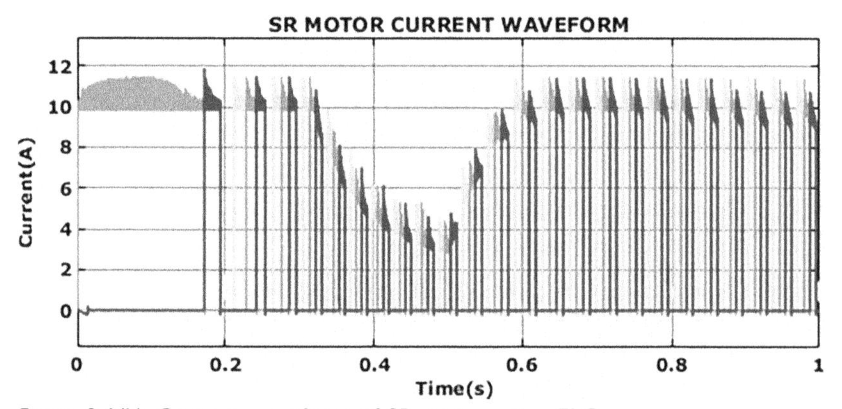

Figure 9.6(b) Current waveform of SR motor using FLC.

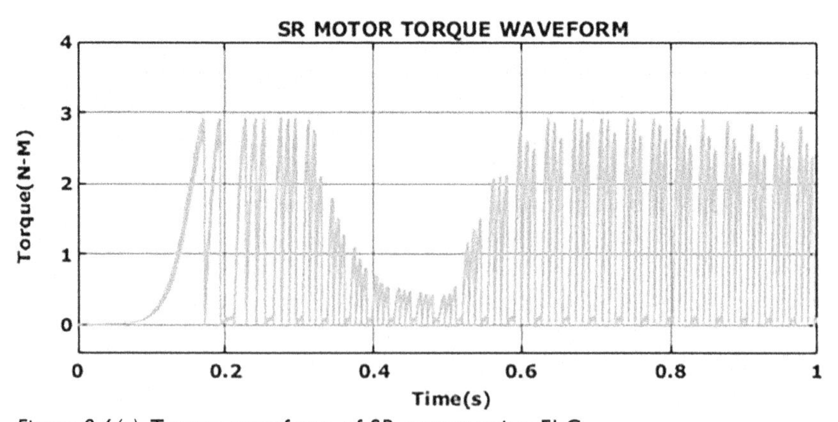

Figure 9.6(c) Torque waveform of SR motor using FLC.

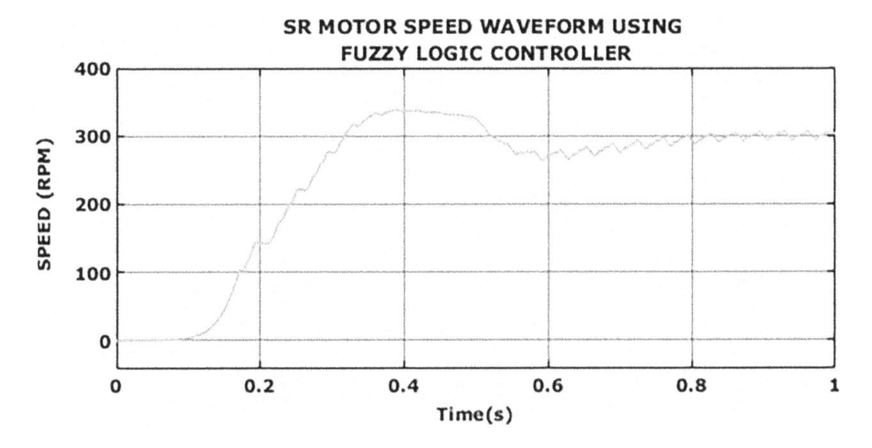

Figure 9.6(d) Speed waveform of SR motor using FLC.

Table 9.5 Performance and Robustness of FL Controller

Parameter	*Tuned*
Rise time (t_{rl}) in Sec	0.36
Peak Time (tp) in Sec	0.37
Overshoot in %	0.48
Settling Time (t_s) in Sec	0.85
Maximum Peak Overshoot	0.51
Steady State error	0.75
Integral Absolute error	9.01
Torque ripple in %	76.14

and longer duration of oscillation. Figure 9.6(d) depicts SRM's speed behaviour when employing FLC.

The performance and robustness of FL controller in terms of time domain parameters is presented in Table 9.5.

The torque ripple in the FLC controller is 76.14% lower than in the PI controller. The performance and robustness of various controllers in terms of time domain parameters is presented in Table 9.6.

From Table 9.6, it can be noted that time domain parameters such as rise time, peak time, overshoot, settling time, and steady state error of FLC are quite reduced when compared to other controllers. The torque ripple is also gradually reduced in the proposed system is shown in Figure 9.7. This illustrates the effectiveness of the proposed scheme in maintaining constant speed of SRM with less settling time with reduced torque ripple.

Table 9.6 Performance and Robustness of Various Controllers

Parameter	*PI*	*FLC*
Rise time (t_{rl}) in Sec	0.32	0.36
Peak Time (tp) in Sec	0.34	0.37
Overshoot in %	0.5	0.48
Settling Time (t_s) in Sec	0.9	0.85
Maximum Peak Overshoot	0.56	0.51
Steady State error	0.82	0.75
Integral Absolute error	16.81	9.01
Torque ripple in %	89.90	76.14

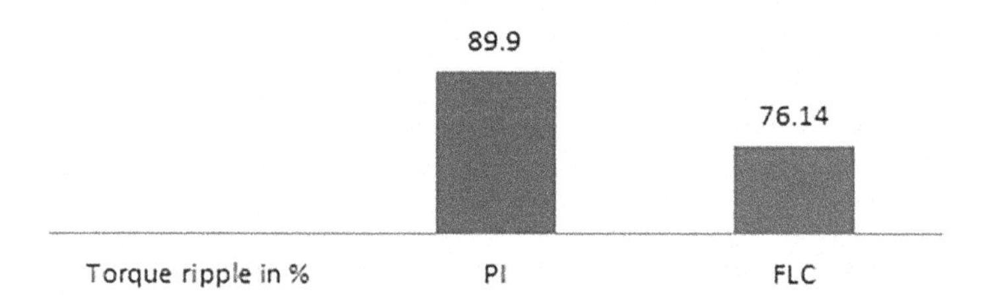

Figure 9.7 Comparison of torque ripple.

9.5 CONCLUSION

This chapter proposes a renewable energy fed SRM powered maritime propulsion system. It validated the SEPIC converter's and the ANFIS-based MPPT algorithm's efficiency in solar PV and battery-powered energy storage systems. The simulation results show that the SEPIC converter has a high voltage gain while being efficient. Renewable energy sources are ideally suited to battery-based energy storage. As a result, an energy storage system powered by batteries has been proposed. Soft computing approaches, such as the fuzzy logic controller, are used to reduce torque ripples in switching reluctance motors. The performance of a typical proportional integral (PI) controller is also evaluated for comparison. For the aforementioned control methods, a comparative analysis in terms of time domain parameters, steady state error, and integral absolute error has been explored. When compared to other controllers, the simulated results reveal that the FLC provides a high efficiency, accurate, and less complex algorithm, as well as improved performance in a dynamic environment. In various scenarios of the suggested model, the torque ripple value and torque improvement percentage are examined. The FLC approach computes the desired output accurately and with less torque. The proposed method has the potential to increase industrial acceptance of SRM for marine propulsion systems.

REFERENCES

[1] O. Abdel-Rahim & H. Wang, 2020, "A new high gain DC-DC converter with model- predictive-control based MPPT technique for photovoltaic systems", *CPSS Transactions on Power Electronics and Applications*, vol. 5, no. 2, pp. 191–200.

[2] Abid Ali Shah Bukhari Belema Prince Alalibo, Wenping Cao, Zhengyu Lin, Shahid Hussain Shaikh, Toufique Ahmed Soomro, Shahzeb Ansar, & Fayyaz Ali Jandan, 2018, "Switched reluctance motor design for electric vehicles based on harmonics and back EMF analysis", *The Journal of Engineering (IET), 9th International Conference on Power Electronics, Machines and Drives*, pp. 1–6.

[3] Ahmad Al Nabulsi & Rached Dhaouadi, 2012, "Efficiency optimization of a DSP-based standalone PV system using fuzzy logic and dual-MPPT control", *IEEE Transactions on Industrial Informatics*, vol. 8, no. 3, pp. 573–584.

[4] S. S. Ahmad & G. Narayanan, 2016, "Linearized modeling of switched reluctance motor for closed-loop current control", *IEEE Transactions on Industry Applications*, vol. 52, no. 4, pp. 3146–3158.

[5] A. Alassi & A. Massoud, 2018, "High-gain DC-DC converters for high power PV applications: Performance assessment", *IEEE 12th International Conference on Compatibility, Power Electronics and Power Engineering*, pp. 1–6.

[6] S. Albert Alexander, M. Srinivasan, S. Ravi, K. Vinoth, S. Senthil Kumar, C. Gnanavel & M. Geetha, 2021, "Fuzzy logic control for solar PV fed modular multilevel inverter towards marine water pumping applications", *IEEE Access*, vol. 9, pp. 88524–88534.

[7] Amresh Kumar Singh, Shailendra Kumar, & Bhim Singh, 2020, "Solar PV energy generation system interfaced to three phase grid with improved power quality", *IEEE Transactions on Industrial Electronics*, vol. 67, no. 5, pp. 3798–3808.

[8] Ankireddy Narendra, Naik N. Venkataramana, Anup Kumar Panda, Nishit Tiwary, & Amit Kumar, 2022, "A single-stage SPV-Fed reduced switching inverter-based sensor less speed control of IM for water pumping applications", *International Transactions on Electrical Energy Systems*, vol. 2022, pp. 1–12.

[9] Atul Sharma, Kamal Kumar Sharma, & Rajesh Chaudhary, 2014, "Torque control in SRM by Neuro-Fuzzy SMC", *International Journal of Research Aspects of Engineering and Management*, vol. 1, no. 2, pp. 42–45.

[10] G. Jegadeeswari, B. Kirubadurai, & D. Lakshmi, 2020, "Multi carrier based new random pulse width modulation for three phase inverters", *International Journal of Scientific and Technology Research (IJSTR)*, vol. 9, no. 2, pp. 5425–5432, ISSN: 2277-8616, impact factor: 3.03.

[11] A. M. Ibrahim, I. Hamdan, L. Nasrat, & M. A. Ismeil, 2022, "Optimal STATCOM for performance improvement of a grid-connected PV/wind hybrid energy system: Case study of Gabal El Zayt Region, Egypt", *SVU-International Journal of Engineering Sciences and Applications*, vol. 3, no. 2, pp. 11–26.

Index

For Product Safety Concerns and Information please contact our EU
representative GPSR@taylorandfrancis.com
Taylor & Francis Verlag GmbH, Kaufingerstraße 24, 80331 München, Germany

www.ingramcontent.com/pod-product-compliance
Ingram Content Group UK Ltd.
Pitfield, Milton Keynes, MK11 3LW, UK
UKHW021120180425
457613UK00005B/160

* 9 7 8 1 0 3 2 6 1 1 9 1 4 *